Math Mammoth
Grade 5 Answer Keys

for the complete curriculum
(Light Blue Series)

Includes answer keys to:

- Worktext part A
- Worktext part B
- Tests
- Cumulative Reviews

By Maria Miller

EDITION 3.0

Contents

Math Mammoth Grade 5-A
Answer Key

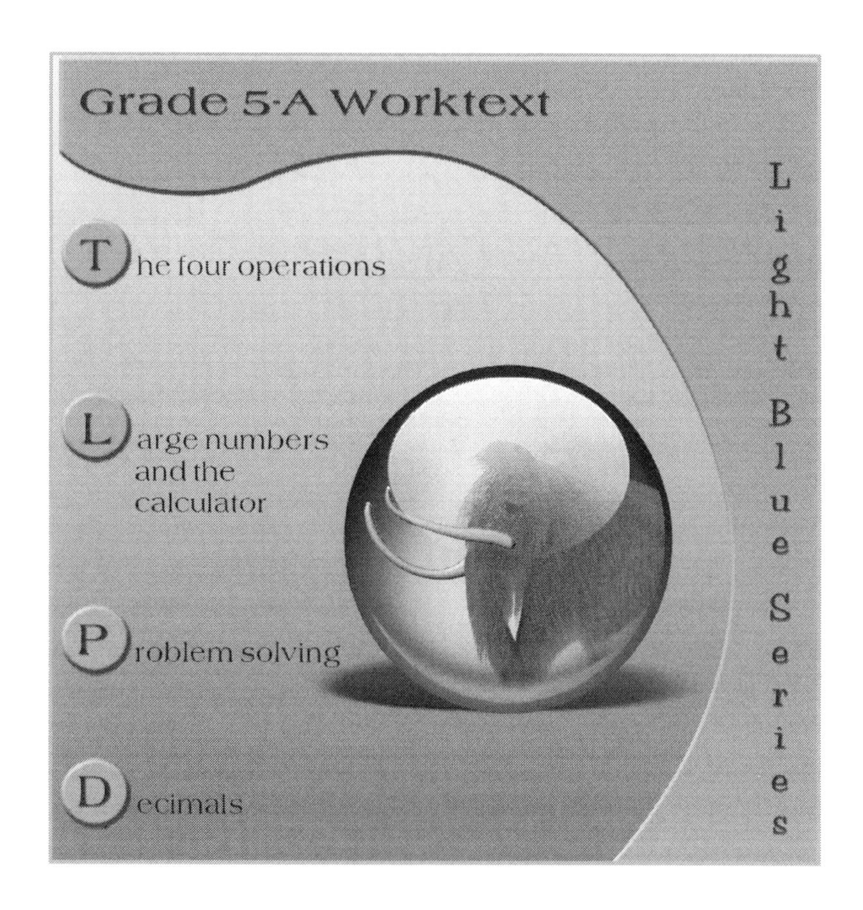

By Maria Miller

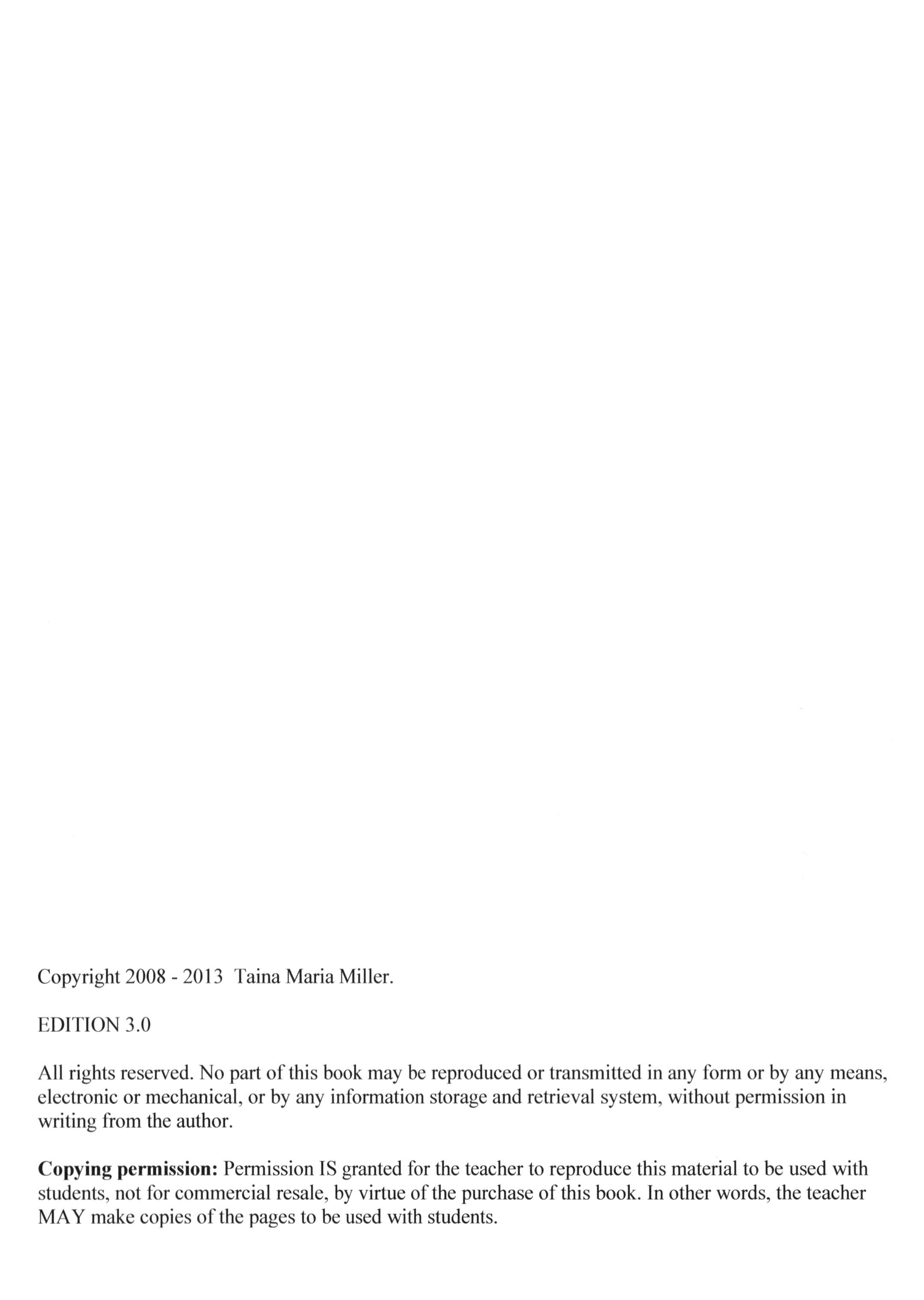

EDITION 3.0

Math Mammoth Grade 5A Answer Key

Contents

Chapter 1: The Four Operations

Warm Up: Mental Math, p. 11 8-19-13

1. a. 38; 75 b. 95; 85 c. 7,800; 17,720
   d. 16; 34 e. 27; 128 f. 1,000; 253

2. The total length of the track is 4 km 300 m.

3. The temperature was 88°F previously.

4. The fourth crate of apples weighs 7 kg.

5. a. 73 b. 210 c. 90

6. See the table below.

7. a. 60; 120; 180; 240; 300; 360; 420; 480; 540
   b. 1,080; 960; 840; 720; 600; 480; 360; 240; 120
   c. 130; 170; 210; 250; 290; 330; 370; 410; 450

8. a. 130 b. 215 c. 246
   d. 535 e. 135 f. 288
   g. 1,435 h. 633 i. 275
   j. 198 k. 128 l. 981

9. a. 150; 900 b. 700; 0 c. 36; 3700
   d. 84; 540 e. 6,300; 380 f. 330; 3,600

6. a. $20 \times 6 = 120$ $200 \times 6 = 1{,}200$ $200 \times 600 = 120{,}000$	b. $10 \times 35 = 350$ $100 \times 35 = 3{,}500$ $20 \times 100 = 2{,}000$	c. $400 \times 500 = 200{,}000$ $60 \times 80 = 4{,}800$ $100 \times 430 = 43{,}000$

The Order of Operations and Equations, p. 13 8-20-13

1. a. 6 b. 260 c. 31 d. 20

2. a. 3 b. 500 c. 63 d. 100

3. a. 68 b. 36 c. 80 d. 44 e. 4 f. 80 g. 108 h. 45

4. a. 18 b. 10 c. 1

5. a. equation b. expression c. equation d. equation e. equation f. expression

6. a. (3) $\$50 - 3 \times \$8 = \$26$. His change was $26.
   b. (2) $6 \times (\$16 - \$5) = \$66$. The total cost is $66.
   c. (4) $(\$8 + \$13) \div 2 = \$10.50$. Andy's share is $10.50
   d. (1) $\$48 \div 4 + \$30 \div 3 = \$22$. Melissa pays $22.

7. a. false b. false c. true. When changing one number in (a) and (b), answers vary. For example:

   a. $1 + \dfrac{32}{8} = 5$ b. $(6 - 2) \times 3 = 5 + 7$

8. a. $(10 + 40 + 40) \times 2 = 180$
   b. $144 = 3 \times (2 + 4) \times 8$
   c. $40 \times 3 = (80 - 50) \times 4$

9.

a. $40 = (11 + 9) \times 2$	b. $4 \times 8 = 5 \times 6 + 2$	c. $4 + 5 = (20 - 2) \div 2$
d. $81 = 9 \times (2 + 7)$	e. $12 \times 11 = 12 + 20 \times 6$	f. $(4 + 5) \times 3 = 54 \div 2$

10. a. $s = 330$ b. $x = 10$ c. $y = 140$

11. Answers will vary. Examples:
    $3 \times 3 + 1 = 1 \times 11 - 1$
    $3 \times 11 + 3 = 3 \times 3 \times 3 + 11 - 1 - 1$
    $11 - 3 = 3 \times 3 - 1$

Review: Addition and Subtraction, p. 16

8-21-13

1. a. $2{,}370 - 1{,}057 = x$ OR $2{,}370 - x = 1{,}057$
   $x + 1{,}057 = 2{,}370$
   Solution: $x = 1{,}313$

   b. $12{,}000 - 3{,}938 - 1{,}506 = x$
   OR $12{,}000 - x - 1{,}506 = 3{,}938$
   OR $12{,}000 - x - 3{,}938 = 1{,}506$
   $3{,}938 + x + 1{,}506 = 12{,}000$
   Solution: $x = 6{,}556$

   c. $560 - 200 = 2x$ OR $560 - 2x = 200$
   $2x + 200 = 560$
   Solution: $x = 180$

2. a. $68 + s$ b. $y - 37$
   c. $60 + b + 40 = 120$ d. $80 - x = 35$

3. a. $20 - (7 + 5)$ b. $20 - 5 - 7$
   c. $20 - (7 - 5)$ d. $(7 - 5) + 20$
   e. $5 + 7 + 20$

4. a. $(15 - 6) + 16$ OR $16 + (15 - 6)$
   b. $100 - (5 + 80)$

5. a. $7{,}000 - (1{,}500 + 2{,}500) = 3{,}000$
   $7{,}000 - 2{,}500 - 1{,}500 = 3{,}000$
   $7{,}000 - (2{,}500 - 1{,}500) = 6{,}000$

   The first and second had the same answer.

   b. $600 + 30 - 30 + 30 - 30 = 600$
   $600 - (30 + 30 + 30 + 30) = 480$
   $600 - 30 - 30 - 30 - 30 = 480$

   The second and third had the same answer.

6. a. yes b. yes c. yes

7. b. $900 − 14 × $58

8. b. 9 × $7 ÷ 2

9. b. $6 \times (26 + 43)$

10. a. The total cost: 15 × $2 + $6 = $36
    Change: $50 − (15 × 2 + 6) = $14

    b. ($9 + $8 + $13) ÷ 3 = $10
    Each child paid $10.

    c. ($128 − 31) × 5 = $485.
    The total cost is $485.

Review: Multiplication and Division, p. 19

1. a. $4 \times 305 = w$; $305 \times 4 = w$
   $w \div 4 = 305$; $w \div 305 = 4$.
   Solution: $w = 1{,}220$

   b. $5 \times w = 305$; $w \times 5 = 305$
   $305 \div w = 5$; $305 \div 5 = w$
   Solution: $w = 61$

2. a. $6 \times y = 90$; $y = 15$
   b. $y \div 6 = 90$; $y = 540$

3. a.
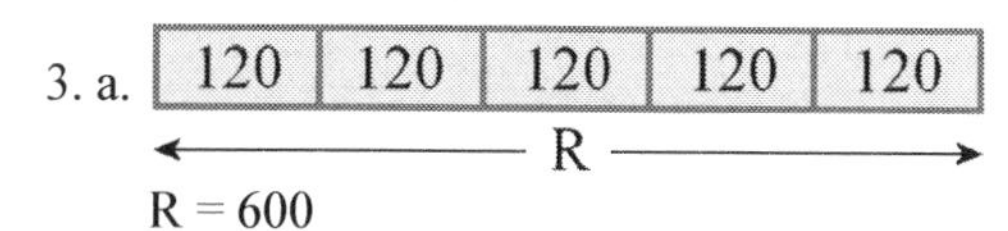

   $R = 600$

   b.
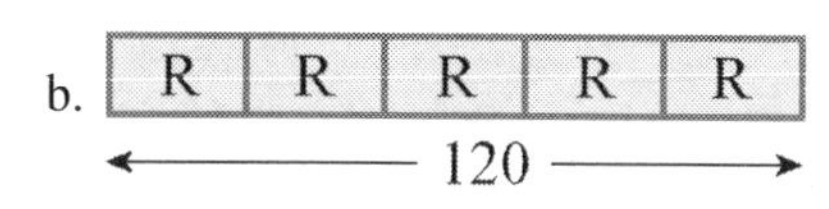

   $R = 24$

   c.
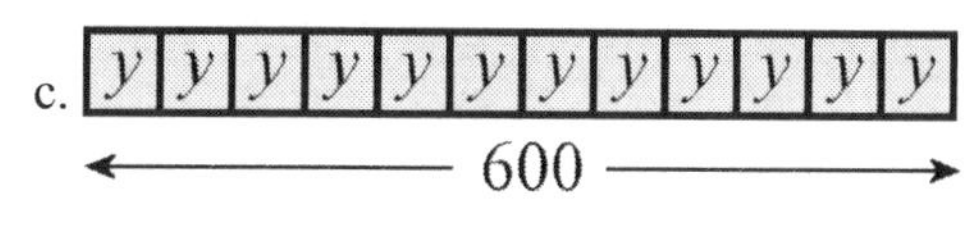

   $y = 50$

   d.
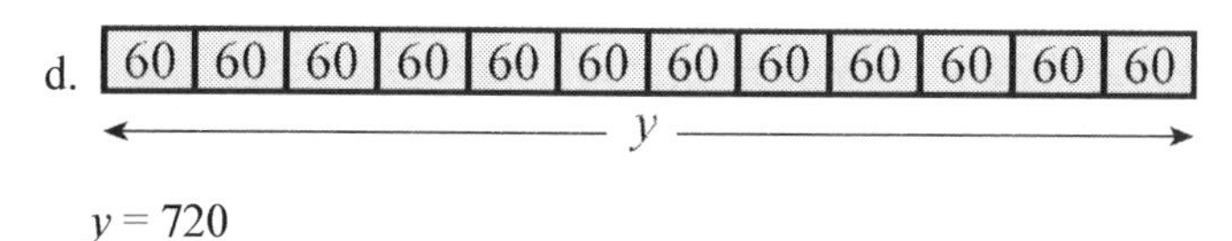

   $y = 720$

4. a. 52×8 b. $15{,}000 \div 300$
   c. $4 \times S \times 18$ d. $80 \div x$
   e. $240 \div 8 = 30$ f. $3 \times 5 \times T = 60$

5. $280 \div N = 4$; $N = 70$

6. $H \div 91 = 3$; $H = 273$

7. a. $30 \times 7 = 210$ b. $12 \times 60 = N$ c. $21 \times 7 = N$
   d. Multiply the quotient by the divisor to find the dividend.

8. a. $450 \div 15 = 30$ $450 \div 30 = 15$
   b. $520 \div 2 = N$ $520 \div N = 2$
   c. $65 \div 5 = N$ $65 \div N = 5$
   d. Divide the product by the known factor to find the value of the unknown factor.

9. a. $72 \div 8 = 9$ b. $350 \div 50 = N$ c. $126 \div 6 = N$
   d. Divide the dividend by the quotient to find the value of the divisor.

10. a. $20 \div 5 = M$; $M = 4$
    b. $3 \times 5 = M$; $M = 15$
    c. $45 \div 5 = M$; $M = 9$
    d. $8{,}800 \div 4 = N$; $N = 2{,}200$
    e. $20 \times 600 = N$; $N = 12{,}000$
    f. $64{,}000 \div 800 = N$; $N = 80$

Multiplying in Parts, p. 23 8-23-13

1.

a. 4 × 27	b. 7 × 83	c. 8 × 56
4 × 20 + 4 × 7	7 × 80 + 7 × 3	8 × 50 + 8 × 6
80 + 28	560 + 21	400 + 48
= 108	= 581	= 448

d. 5 × 216	e. 4 × 3,481
5 × 200 + 5 × 10 + 5 × 6	4 × 3,000 + 4 × 400 + 4 × 80 + 4 × 1
1,000 + 50 + 30	12,000 + 1,600 + 320 + 4
= 1,080	= 13,924

2.

a.

		4 9 2
		× 6
6 × 2 →		1 2
6 × 90 →		5 4 0
6 × 400 →	+	2 4 0 0
		2 9 5 2

b.

	2 5 5
	× 4
	2 0
	2 0 0
+	8 0 0
	1 0 2 0

c.

	8 1 7
	× 7
	4 9
	7 0
+	5 6 0 0
	5 7 1 9

3.

a.

	2 5 1 0
	× 9
9 × 0 →	0
9 × 10 →	9 0
9 × 500 →	4 5 0 0
9 × 2,000 →	+ 1 8 0 0 0
	2 2 5 9 0

b.

4 4 7 8
× 5
4 0
3 5 0
2 0 0 0
+ 2 0 0 0 0
2 2 3 9 0

c.

2 6 0 7 2
× 6
1 2
4 2 0
0
3 6 0 0 0
+ 1 2 0 0 0 0
1 5 6 4 3 2

4.

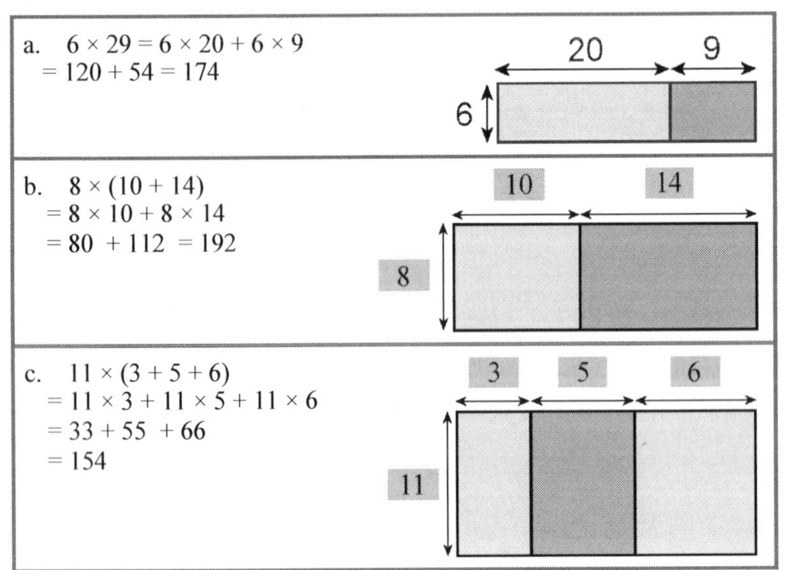

Multiplying in Parts, cont. 8-26-13

5.

a.

28 × 16 = 20 × 16 and 8 × 16

20 × 10 and 20 × 6 and 8 × 10 and 8 × 6

200 and 120 and 80 and 48

Now add the parts to get the total: 448

b.

71 × 25 = 70 × 25 and 1 × 25

70 × 20 and 70 × 5 and 1 × 20 and 1 × 5

1400 and 350 and 20 and 5

Now add the parts to get the total: 1775

c.

48 × 19 = 40 × 19 and 8 × 19

40 × 10 and 40 × 9 and 8 × 10 and 8 × 9

400 and 360 and 80 and 72

Now add the parts to get the total: 912

d.

39 × 94 = 30 × 94 and 9 × 94

30 × 90 and 30 × 4 and 9 × 90 and 9 × 4

2700 and 120 and 810 and 36

Now add the parts to get the total: 3,666

6.

a.		b.	
	8 7		2 4
	× 1 5		× 7 1
5 × 7 →	3 5	1 × 4 →	4
5 × 80 →	4 0 0	1 × 20 →	2 0
10 × 7 →	7 0	70 × 4 →	2 8 0
10 × 80 →	+ 8 0 0	70 × 20 →	+ 1 4 0 0
	1 3 0 5		1 7 0 4

7. a. You can find the area of the green rectangle by subtracting: 153 − 117 = 36. The green rectangle has an area of 36 square units.
   b. The missing side lengths are 13 units and 9 units. Once you know the green rectangle has an area of 36 square units, then its other side must be 9 units (because 9 × 4 = 36). Then, knowing the yellow rectangle is 117 square units, and its one side is 9 units, you can find its other side by dividing: 117 ÷ 9 = 13.

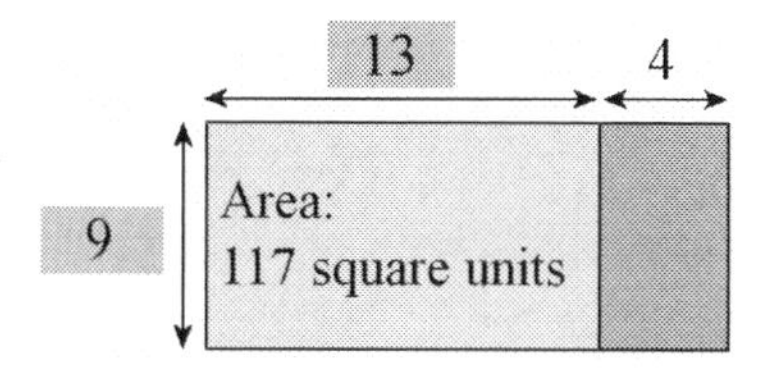

8. c. 26 × $18 + 26 × $8 and d. 26 × ($18+ $8)

9. a. Not the same. b. Yes, the same. c. No, not the same.
   d. Yes, the same. e. No, not the same. f. Yes, the same.

10.

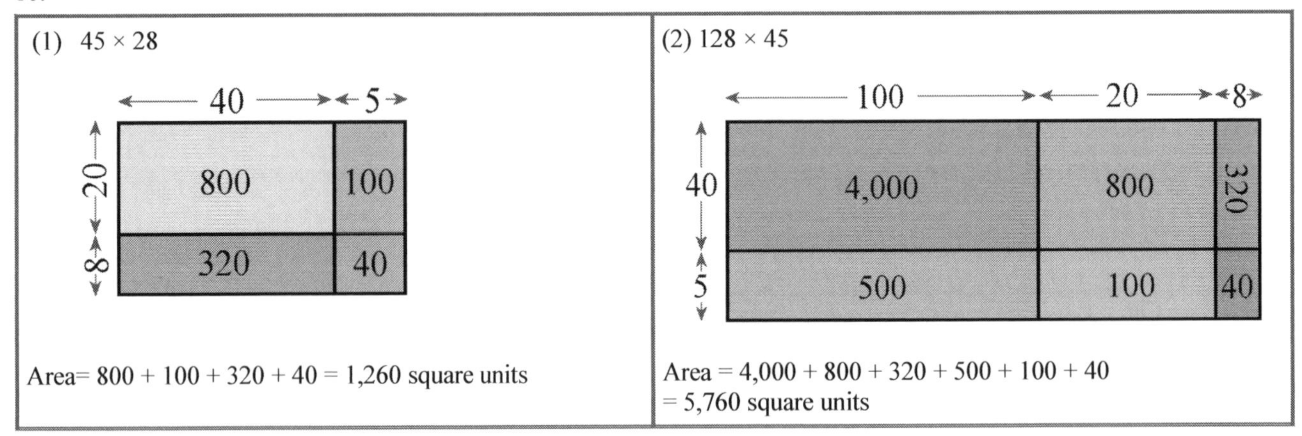

11. a. Draw a rectangle with sides 41 and 63. Divide the short side into two lengths: the tens, "40," and the units, "1." Divide the long side similarly into "60" and "3." Now you have divided the large rectangle into four smaller rectangles. (The image on the right is not to scale.)

	60	3
40	2,400	120
1	60	3

b. Show Michael that his "method" solves for only *two* of the four smaller rectangles: "40 × 60" and "1 × 3." There are still two rectangles left: "1 × 60" and "40 × 3."
41 × 63 can be broken down into tens and ones, but you have to multiply the tens and ones from *both* numbers. The method that Michael suggested does not include the tens and ones from both numbers. To multiply correctly, we have to multiply 40 × 60 and 1 × 60, then multiply 40 × 3 and 1 × 3, then add all of these products:
40 × 60 = 2,400; 1 × 60 = 60; 40 × 3 = 120; 1 × 3 = 3; then 2,400 + 60 + 120 + 3 = 2,583—which is 180 more than Michael's method of multiplying.

Puzzle corner:
143 × 715 = 100 × 700 + 40 × 700 + 3 × 700 + 100 × 10 + 40 × 10 + 3 × 10 + 100 × 5 + 40 × 5 + 3 × 5
= 70,000 + 28,000 + 2,100 + 1,000 + 400 + 30 + 500 + 200 + 15 = 102,245

The image below is not to scale.

	700	10	5
100	70,000	1,000	500
40	28,000	400	200
3	2,100	30	15

The Multiplication Algorithm, p. 29

8-27-13

1. a. 280 b. 3,320 c. 1,172 d. 7,016
   e. 12,264 f. 10,460 g. 26,169 h. 71,368

2. a. 122,864 b. 83,415 c. 720,615 d. 812,442

3. Estimates will vary.
   a. Estimate: $5 \times 9{,}000 = 45{,}000$; Exact: 44,355
   b. Estimate: $4 \times 22{,}000 = 88{,}000$; Exact: 89,596
   c. Estimate: $7 \times 90{,}000 = 630{,}000$; Exact: 610,680
   d. Estimate: $4 \times 212{,}000 = 848{,}000$; Exact: 851,152
   e. Estimate: $6 \times 55{,}000 = 330{,}000$; Exact: 329,532
   f. Estimate: $3 \times 170{,}000 = 510{,}000$; Exact: 519,117

4. a. \$181.76 b. \$326.00 c. \$2,071.50 d. \$3,564.50

5. Estimates will vary.
   a. Estimate: $2 \times \$57 = \$114$; Exact: \$113.10
   b. Estimate: $6 \times \$130 = \$780$; Exact: \$773.70
   c. Estimate: $\$100 - 3 \times \$32 = \$4$; Exact: \$4.75

6. Estimates will vary.
   a. Estimate: $90 \times 30 = 2{,}700$; Exact: 2.511
   b. Estimate: $50 \times 50 = 2{,}500$; Exact: 2,530
   c. Estimate: $80 \times 20 = 1{,}600$; Exact: 1,392
   d. Estimate: $60 \times 90 = 5{,}400$; Exact: 5,490
   e. Estimate: $20 \times 20 = 400$; Exact: 432
   f. Estimate: $100 \times 50 = 5{,}000$; Exact: 4,998

7. Estimates will vary.
   a. Estimate: $200 \times 25 = 5{,}000$; Exact: 4,775
   b. Estimate: $220 \times 40 = 8{,}800$; Exact: 9,184
   c. Estimate: $600 \times 40 = 24{,}000$; Exact: 21,035
   d. Estimate: $250 \times 80 = 20{,}000$; Exact: 19,890
   e. Estimate: $200 \times 10 = 2{,}000$; Exact: 2,167
   f. Estimate: $300 \times 40 = 12{,}000$; Exact: 11,340

More Multiplication, p. 34

8-28-13

1. a. 46,795 b. 93,252 c. 33,702
   d. 589,245 e. 166,221 f. 82,750

2. Estimates will vary.
   a. Estimate: $1{,}400 \times 30 = 42{,}000$; Exact: 40,257
   b. Estimate: $2{,}000 \times 35 = 70{,}000$; Exact: 72,975
   c. Estimate: $60{,}000 \times 50 = 3{,}000{,}000$; Exact: 2,821,680
   d. Estimate: $90{,}000 \times 80 = 7{,}200{,}000$; Exact: 7,422,312
   e. Estimate: $2{,}000 \times 400 = 800{,}000$; Exact: 816,864
   f. Estimate: $6{,}000 \times 300 = 1{,}800{,}000$; Exact: 1,857,744

3. a. 100,000 b. 6,300 c. 10,000 d. 800,000
   e. 960,000 f. 60,000 g. 3,200,000 h. 1,200,000

4.

a. $500 \times 29 = 14{,}500$ Simply multiply 5×29, then tag two zeros on the final answer.	b. $340 \times 210 = 71{,}400$ Multiply 34×21, then tag two zeros on the final answer.	c. $280 \times 700 = 196{,}000$ Multiply 28×7, then tag three zeros to the final answer.
d. $99 \times 9{,}900 = 980{,}100$	e. $600 \times 1{,}800 = 1{,}080{,}000$	f. $24{,}500 \times 30 = 735{,}000$

More Multiplication, cont.

5.

a. Estimate: 1,700 − 5 × 140 = 1,000. Exact: 1,059
b. Estimate: 2 × \$27.00 + 3 × \$3.00 = \$63.00. Exact: \$63.77
c. Estimate: 370 × 20 = 7,400. Exact: 8,760 hours in a year.
d. Estimate: 15,000 − 160 × 30 = 10,200. Exact: 10,040 lb left
e. Estimate: \$120 × 50 = \$6,000. Exact: \$6,240 in savings each year.

6. a. 3 × 365 + 366 = 1,461 days.
   b. 4 × 365 + 366 = 1,826 days.
   c. Answers will vary.

Puzzle corner: a. 9,628,557 b. 50,611,995

Long Division, p. 39

1. a. 294; 294 × 7 = 2,058 b. 437; 437 × 9 = 3,933
   c. 547; 547 × 6 = 3,282 d. 689; 689 × 6 = 4,134

2. a. 1,078; Check: 1,078 × 7 = 7,546 b. 2,406; Check: 2,406 × 3 = 7,218
   c. 1,908; Check: 1,908 × 4 = 7,632 d. 516 ÷ 6 = 86, 516 − 86 = 430. He has \$430 left to spend.

3. a. 2,340 b. 515 c. 810 d. 303 e. 505 f. 306 g. 504 h. 3,205 i. 1,030

4. \$2,440 ÷ 4 = \$610, \$2,440 − \$610 = \$1,830. The vacation costs now \$1,830.

5. 2,100 mi ÷ 5 = 420 mi, 420 mi × 4 = 1,680 mi. There are 1,680 miles left to travel.

6. 6 × 117 + 3 = 705, so, no the division is not correct.

7. a. There are 3 feet in one yard.
   b. 227 ÷ 3 = 75 R2, so 227 ft = 75 yd 2 ft.

8. 80 ÷ 3 = 26 R2. Sally got 26 full bags (and one bag with 2 kg of flour).

9. 911 ÷ 5 = 182 R1, 546 ÷ 5 = 109 R1, 77 ÷ 5 = 15 R 2
   The shortcut is: divide the last digit (or even the last two digits) of the number by 5, and find the remainder of that division.
   For example, in 911 ÷ 5, simply check the remainder of the division 1 ÷ 5 (or 11 ÷ 5). It is 1.
   Or, 77 ÷ 5: simply check what the remainder is when dividing 7 by 5. It is 2.

Puzzle corner. First, convert the liquid ounces into cups and ounces: 473 ÷ 8 = 59 R1, so 473 oz = 59 C 1 oz. Then, convert the 59 cups into quarts and cups: 59 C = 14 qt 3 C. Lastly, convert the 14 quarts into gallons and quarts: 14 qt = 3 gal 2 qt. Then gather all the parts together to obtain 473 oz = 3 gal 2 qt 3 C 1 oz.

A Two-Digit Divisor 1, p. 43

1.

Table of 21:
2 × 21 = 42
3 × 21 = 63
4 × 21 = 84
5 × 21 = 105
6 × 21 = 126
7 × 21 = 147
8 × 21 = 168
9 × 21 = 189

```
       1 8 2
21 ) 3 8 2 2
     2 1
     1 7 2
     1 6 8
         4 2
         4 2
           0
```

Check:

```
    1 8 2
  ×   2 1
  -------
    1 8 2
  3 6 4 0
  -------
  3 8 2 2
```

2.

a. Table of 15:
2 × 15 = 30
3 × 15 = 45
4 × 15 = 60
5 × 15 = 75
6 × 15 = 90
7 × 15 = 105
8 × 15 = 120
9 × 15 = 135

```
       3 2 1
15 ) 4 8 1 5
     4 5
       3 1
       3 0
         1 5
         1 5
```

Check:

```
    3 2 1
  ×   1 5
  -------
  1 6 0 5
  3 2 1 0
  -------
  4 8 1 5
```

b. Table of 12:
2 × 12 = 24
3 × 12 = 36
4 × 12 = 48
5 × 12 = 60
6 × 12 = 72
7 × 12 = 84
8 × 12 = 96
9 × 12 = 108

```
       4 2 9
12 ) 5 1 4 8
     4 8
       3 4
       2 4
       1 0 8
       1 0 8
           0
```

Check:

```
    4 2 9
  ×   1 2
  -------
    8 5 8
  4 2 9 0
  -------
  5 1 4 8
```

c. Table of 25:
2 × 25 = 50
3 × 25 = 75
4 × 25 = 100
5 × 25 = 125
6 × 25 = 150
7 × 25 = 175
8 × 25 = 200
9 × 25 = 225

```
       2 5 1
25 ) 6 2 7 5
     5 0
     1 2 7
     1 2 5
         2 5
         2 5
           0
```

Check:

```
    2 5 1
  ×   2 5
  -------
  1 2 5 5
  5 0 2 0
  -------
  6 2 7 5
```

d. Table of 16:
2 × 16 = 32
3 × 16 = 48
4 × 16 = 64
5 × 16 = 80
6 × 16 = 96
7 × 16 = 112
8 × 16 = 128
9 × 16 = 144

```
         9 4
16 ) 1 5 0 4
     1 4 4
         6 4
         6 4
           0
```

Check:

```
      9 4
  ×   1 6
  -------
    5 6 4
    9 4 0
  -------
  1 5 0 4
```

3.

a.

Table of 12:
2 × 12 = 24
3 × 12 = 36
4 × 12 = 48
5 × 12 = 60
6 × 12 = 72
7 × 12 = 84
8 × 12 = 96
9 × 12 = 108

```
       7 4
 12 ) 8 8 8
      8 4
        4 8
        4 8
          0
```

Check:

```
    7 4
  × 1 2
  -----
  1 4 8
  7 4 0
  -----
  8 8 8
```

b.

Table of 22:
2 × 22 = 44
3 × 22 = 66
4 × 22 = 88
5 × 22 = 110
6 × 22 = 132
7 × 22 = 154
8 × 22 = 176
9 × 22 = 198

```
         3 0 5
 22 ) 6 7 1 0
      6 6
        1 1 0
        1 1 0
            0
```

Check:

```
    3 0 5
  ×   2 2
  -------
    6 1 0
  6 1 0 0
  -------
  6 7 1 0
```

c.

Table of 14:
2 × 14 = 28
3 × 14 = 42
4 × 14 = 56
5 × 14 = 70
6 × 14 = 84
7 × 14 = 98
8 × 14 = 112
9 × 14 = 126

```
         1 2 4
 14 ) 1 7 3 6
      1 4
        3 3
        2 8
          5 6
          5 6
            0
```

Check:

```
    1 2 4
  ×   1 4
  -------
    4 9 6
  1 2 4 0
  -------
  1 7 3 6
```

d.

Table of 51:
2 × 51 = 102
3 × 51 = 153
4 × 51 = 204
5 × 51 = 255
6 × 51 = 306
7 × 51 = 357
8 × 51 = 408
9 × 51 = 459

```
         1 4 8
 51 ) 7 5 4 8
      5 1
      2 4 4
      2 0 4
        4 0 8
        4 0 8
            0
```

Check:

```
    1 4 8
  ×   5 1
  -------
    1 4 8
  7 4 0 0
  -------
  7 5 4 8
```

4. a. 40; 41 b. 14; 15 c. 10; 20
d. 12; 13 e. 20; 23 f. 20; 22

5. a. There are 12 inches in 1 foot.
b. 245 in. = 20 ft 5 in.
c. 387 in. = 32 ft 3 in.

6. a. There are 16 ounces in 1 pound.
b. 163 oz = 10 lb 3 oz.
c. 473 oz = 29 lb 9 oz.

7. The baby would gain 365 ounces. 365 ÷ 16 = 22 R13, so 365 oz = 22 lb 13 oz. The baby would gain 22 lb 13 oz.

A Two-Digit Divisor 2, p. 47

1.

2 × 37 = 74 3 × 37 = 111 4 × 37 = 148 5 × 37 = 185 6 × 37 = 222 7 × 37 = 259 8 × 37 = 296 9 × 37 = 333	1 1 1 a. 37) 4 1 0 7 -3 7 4 0 -3 7 3 7 - 3 7 0	1 1 1 × 3 7 7 7 7 3 3 3 0 4 1 0 7
2 × 58 = 116 3 × 58 = 174 4 × 58 = 232 5 × 58 = 290 6 × 58 = 348 7 × 58 = 406 8 × 58 = 464 9 × 58 = 522	7 6 b. 58) 4 4 0 8 -4 0 6 3 4 8 -3 4 8 0	7 6 × 5 8 6 0 8 3 8 0 0 4 4 0 8
2 × 96 = 192 3 × 96 = 288 4 × 96 = 384 5 × 96 = 480 6 × 96 = 576 7 × 96 = 672 8 × 96 = 768 9 × 96 = 864	1 0 2 c. 96) 9 7 9 2 -9 6 1 9 2 -1 9 2 0	1 0 2 × 9 6 6 1 2 9 1 8 0 9 7 9 2

2.

2 × 48 = 96 3 × 48 = 144 4 × 48 = 192 5 × 48 = 240 6 × 48 = 288 7 × 48 = 336 8 × 48 = 384 9 × 48 = 432	1 2 5 a. 48) 6 0 1 1 -4 8 1 2 1 - 9 6 2 5 1 - 2 4 0 1 1	1 2 5 × 4 8 1 0 0 0 + 5 0 0 0 6 0 0 0 + 1 1 6 0 1 1
2 × 92 = 184 3 × 92 = 276 4 × 92 = 368 5 × 92 = 460 6 × 92 = 552 7 × 92 = 644 8 × 92 = 736 9 × 92 = 828	9 4 b. 92) 8 7 1 2 -8 2 8 4 3 2 - 3 6 8 6 4	9 4 × 9 2 1 8 8 + 8 4 6 0 8 6 4 8 + 6 4 8 7 1 2
2 × 55 = 110 3 × 55 = 165 4 × 55 = 220 5 × 55 = 275 6 × 55 = 330 7 × 55 = 385 8 × 55 = 440 9 × 55 = 495	1 2 2 c. 55) 6 7 4 5 -5 5 1 2 4 1 1 0 1 4 5 - 1 1 0 3 5	1 2 2 × 5 5 6 1 0 + 6 1 0 0 6 7 1 0 + 3 5 6 7 4 5

A Two-Digit Divisor 2, cont.

3. 3 × $156 = $468, $468 ÷ 12 = $39. One payment is $39.

4. a. $3600 ÷ 15 = $240, $240 × 4 = $960. He paid $960 in taxes.
   b. $3600 − $960 = $2,640. He has $2,640 left.

5.

			b. 3	5				
a. 5	1		9			d. 2	7	f. 6
7			f. 5	0	8	1		0
	c. 8	4	0			0		9
	3			e. 4	2	3		
	0			0				
				8				

Puzzle corner: a. 12,408 ÷ 118 = 105 R18 b. 70,854 ÷ 235 = 301 R119

Long Division and Repeated Subtraction, p. 50

1.

a. Bag 657 apples;
3 apples in each bag.

Apples	Bags
657	
− 300	100
357	
− 300	100
57	
− 30	10
27	
− 27	9
0	219

b. Bag 984 peaches;
8 in each bag.

Peaches	Bags
984	
− 800	100
184	
− 80	10
104	
− 80	10
24	
− 24	3
0	123

c. Bag 536 pineapples;
4 in each bag.

Pineapples	Bags
536	
− 400	100
136	
− 40	10
96	
− 40	10
56	
− 40	10
16	
− 16	4
0	134

2.

a. Bag 474 apples; 3 apples in each bag.

Apples	Bags
474	
− 300	100
174	
− 130	50
24	
− 24	8
0	158

b. Bag 2,032 lemons; 8 lemons in each bag.

Lemons	Bags
2032	
− 1600	200
432	
− 400	50
32	
− 32	4
0	254

c. Bag 3,655 bananas; 5 in each bag.

Bananas	Bags
3655	
− 3500	700
155	
− 150	30
5	
− 5	1
0	731

d. Bag 762 mangos; 6 mangos in each bag.

Mangos	Bags
762	
− 600	100
162	
− 120	20
42	
− 42	7
0	127

e. Bag 1,152 papayas; 3 papayas in each bag.

Papayas	Bags
1152	
− 900	300
252	
− 240	80
12	
− 12	4
0	384

f. Bag 4,770 cherries; 9 in each bag.

Cherries	Bags
4770	
− 4500	500
270	
− 270	30
0	530

3. There would still be 127 bags, but there would be a remainder of 3 mangos.

4. a. 168 b. 47

The last two examples of the teaching box before problem 5:

Dividend (the apples)	Quotient (the bags)
988	
− 800	200
188	
− 160	40
28	
− 28	7
0	247

```
     2 4 7
4 ) 9 8 8
  - 8 0 0
    1 8 8
  - 1 6 0
      2 8
    - 2 8
        0
```

Dividend (the apples)	Quotient (the bags)
2546	
− 2100	300
446	
− 420	60
26	
− 21	3
5	363

```
      3 6 3
7 ) 2 5 4 6
  - 2 1 0 0
      4 4 6
  -   4 2 0
        2 6
      - 2 1
          5
```

5.

a. Bag 610 apples, 5 apples in each bag.

Apples	Bags
610	
− 500	100
110	
− 100	20
10	
− 10	2
0	122

```
     1 2 2
5 ) 6 1 0
   -5
    1 1
   -1 0
      1 0
     -1 0
        0
```

b. Bag 853 kiwis, 3 kiwis in each bag.

Kiwis	Bags
853	
− 600	200
253	
− 240	80
13	
− 12	4
1	284

```
     2 8 4
3 ) 8 5 3
   -6
    2 5
   -2 4
      1 3
    - 1 2
        1
```

c. Bag 445 grapefruits, 3 grapefruits in each bag.

Grapefruits	Bags
445	
− 300	100
145	
− 120	40
25	
− 24	8
1	148

```
     1 4 8
3 ) 4 4 5
   -3
    1 4
   -1 2
      2 5
    - 2 4
        1
```

d. Bag 952 plums, 4 plums in each bag.

Plums	Bags
952	
− 800	200
152	
− 120	30
32	
− 32	8
0	238

```
     2 3 8
4 ) 9 5 2
   -8
    1 5
   -1 2
      3 2
    - 3 2
        0
```

e. Bag 2,450 pears, 9 pears in each bag.

Pears	Bags
2450	
− 1800	200
650	
− 630	70
20	
− 18	2
2	272

```
       2 7 2
9 ) 2 4 5 0
   -1 8
      6 5
    - 6 3
        2 0
      - 1 8
          2
```

f. Bag 1,496 oranges, 8 oranges in each bag.

Oranges	Bags
1496	
− 800	100
696	
− 640	80
56	
− 56	7
0	187

```
       1 8 7
8 ) 1 4 9 6
    - 8
      6 9
    - 6 4
        5 6
      - 5 6
          0
```

Divisibility Rules, p. 55

1. a. No, because 8 does not divide evenly into 100.
   b. Yes, because 3,500 divided by 7 is an even division.
   c. No, because 50 divided by 9 leaves a remainder (is not an even division).

2. Answers vary.
   a. No, because 283 ÷ 13 leaves a remainder.
   b. Yes, because 13 is a factor of the number (13 × 2,809). The number (13 × 2,809) is divisible by 13 because it is 13 times some number.
   c. No. When you multiply odd numbers, the product is also odd, so it cannot be divisible by 2.
   d. No, because 9,896 ÷ 7 leaves a remainder.
   e. Yes, because 2 × 5 equals 10, so this number is actually (10 × 758) or 7,580, which naturally is divisible by 10.
   f. Yes. Since 2 × 2 = 4, this number is (4 × 15 × 7), or 4 times some number. So, the number must be divisible by 4.

3.

Divisible by	2	5	10	100	1000
825		X			
400	X	X	X	X	
332	X				

Divisible by	2	5	10	100	1000
600,200	X	X	X	X	
56,000	X	X	X	X	X
307,995		X			

4. a. No. b. Yes. 43,719 ÷ 3 = 14,573 c. No

5. Change the last 2 to a 1. 238,881 ÷ 3 = 79,627

6. Mystery Number: 84; 132

7. a. No b. Yes. 576 ÷ 9 = 64 c. Yes. 44,082 ÷ 9 = 4,898.

8.

Divisible by	2	3	5	6	9
589					
558	X	X		X	X

Divisible by	2	3	5	6	9
495		X	X		X
3,594	X	X		X	

9. a. 99, 108, 117, 126, 135 or 99, 90, 81, 72, 63 b. 686, 679, 672, 665, 658

10.

Divisible by	2	3	4	5	6	9
1,755		X		X		X
298	X					
4,000	X		X	X		
3,270	X	X		X	X	

Divisible by	2	3	4	5	6	9
3,548	X		X			
277						
237		X				
10,999						

11.

Divisible by	2	3	4	5	6	8	9
628	X		X				
405		X		X			X

Divisible by	2	3	4	5	6	8	9
938	X						
224	X		X			X	

12.

a. 26 ÷ 4 = 6 R2 27 ÷ 4 = 6 R3 28 ÷ 4 = 7 R0 29 ÷ 4 = 7 R1 30 ÷ 4 = 7 R2 31 ÷ 4 = 7 R3 32 ÷ 4 = 8 R0	b. 78 ÷ 3 = 26 R0 79 ÷ 3 = 26 R1 80 ÷ 3 = 26 R2 81 ÷ 3 = 27 R0 82 ÷ 3 = 27 R1 83 ÷ 3 = 27 R2 84 ÷ 3 = 28 R0	c. 54 ÷ 7 = 7 R5 55 ÷ 7 = 7 R6 56 ÷ 7 = 8 R0 57 ÷ 7 = 8 R1 58 ÷ 7 = 8 R2 59 ÷ 7 = 8 R3 60 ÷ 7 = 8 R4

13. a. The remainder is 1. b. 2 c. 3

14. a. The remainder is 1. b. 5 c. 9

15. a. 96 b. 97

Divisibility Rules, cont.

16. **Divisible by 4:**

18	52	100	502	300	312	348	322
16	44	64	446	292	144	360	422
6	16	72	292	280	266	436	232
86	94	104	144	216	204	568	522
60	54	128	132	244	286	572	588
12	8	12	90	308	312	78	544
15	12	136	98	254	308	348	548
44	48	66	166	256	388	428	444

Divisible by 3:

5	15	23	392	486	500	510	581
3	9	14	298	471	492	501	555
6	21	35	255	444	504	398	577
15	27	39	65	408	354	345	362
17	37	41	99	103	287	285	311
21	33	44	81	88	204	234	254
22	36	51	69	127	171	202	189
9	16	33	72	108	132	156	166

17. Mystery Number: 32; 210

Review: Factors and Primes, p. 60

1. Answers may vary. For example:

product	factors
a. 10	5 × 2
b. 50	5 × 10

product	factors
c. 120	2 × 5 × 12
d. 22	2 × 11

product	factors
e. 54	6 × 9
f. 72	2 × 4 × 9

2.

Number	Divisible by:													
	1	2	3	4	5	6	7	8	9	10	11	12	13	14
8	X	X		X				X						
9	X		X						X					
10	X	X			X					X				
11	X										X			
12	X	X	X	X		X						X		
13	X												X	
14	X	X					X							X

Review: Factors and Primes, cont.

3. 2, 3, 5, 7, 11 and 13 are prime numbers

4. Primes: 17, 19, 23, 29, 31, 37, 41, 43, 47

number	divisible by									
	1	2	3	4	5	6	7	8	9	10
15	X		X		X					
16	X	X		X				X		
17	X									
18	X	X	X			X			X	
19	X									
20	X	X		X	X					X
21	X		X				X			
22	X	X								
23	X									
24	X	X	X	X		X		X		
25	X				X					
26	X	X								
27	X		X						X	
28	X	X		X			X			
29	X									
30	X	X	X		X	X				X
31	X									
32	X	X		X				X		
33	X		X							
34	X	X								
35	X				X		X			
36	X	X	X	X		X			X	
37	X									
38	X	X								
39	X		X							
40	X	X		X	X			X		X
41	X									
42	X	X				X	X			
43	X									
44	X	X		X						
45	X		X		X				X	
46	X	X								
47	X									

Review: Factors and Primes, cont.

5. a. 1, 2, 13, 26 b. 1, 2, 19, 38
   c. 1, 2, 4, 8, 11, 22, 44, 88 d. 1, 47
   e. 1, 71 f. 1, 2, 43, 86

6. a. 1, 2, 3, 4, 6, 8, 12, 24 b. 1, 2, 29, 58
   c. 1, 2, 4, 8, 16, 32, 64 d. 1, 2, 3, 4, 6, 8, 12, 16, 24, 32, 48, 96

7. a. 83 b. 78 c. 132 or 138

Puzzle corner: a. 113 or 119 b. 36 c. 30

Prime Factorization, p. 64

1. a. 18 = 2 × 3 × 3 b. 6 = 2 × 3 c. 14 = 2 × 7
   d. 8 = 2 × 2 × 2 e. 12 = 2 × 2 × 3 f. 20 = 2 × 2 × 5
   g. 16 = 2 × 2 × 2 × 2 h. 24 = 2 × 2 × 2 × 3 i. 27 = 3 × 3 × 3
   j. 25 = 5 × 5 k. 33 = 3 × 11 l. 15 = 3 × 5

2. a. 42 = 2 × 3 × 7 b. 56 = 2 × 2 × 2 × 7 c. 68 = 2 × 2 × 17
   d. 75 = 3 × 5 × 5 e. 47 = 1 × 47 f. 99 = 3 × 3 × 11
   g. 72 = 2 × 2 × 2 × 3 × 3 h. 80 = 2 × 2 × 2 × 2 × 5 i. 97 = 1 × 97
   j. 85 = 5 × 17 k. 66 = 2 × 3 × 11 l. 82 = 2 × 41

3. a. 110 b. 24 c. 42 d. 66 e. 90 f. 102

4. a. 130 b. 2,002 c. 570

5. Answers will vary. Please check the students' work.

Puzzle corner: a. 2,145 = 3 × 5 × 11 × 13 b. 3,680 = 2 × 2 × 2 × 2 × 2 × 5 × 23 c. 10,164 = 2 × 2 × 3 × 7 × 11 × 11

Chapter 1 Review, p. 69

1. a. 281 b. 69 c. 95,118

2. 83,493 – 21,390 = 62,103

3. a. 55 b. 140 c. 30 d. 56

4. a. 606 b. 902 c. 810 d. 93 e. 1,201

5. a. 9 b. 3 c. 8

6. a. $x - 9$ b. $y + 3 + 8 = 28$ c. $60 \div b = 12$ d. $8 \times x \times y$

7. **(4)** 4 × $3.75 ÷ 3 = $5. Each girl paid $5.

8. a. (12 + 17) ÷ 2 = $14.50. Each paid $14.50.
   b. 5 × 4.50 – 2 = $20.50. Henry paid $20.50.

9. a. R ÷ 4 = 544; R = 2,176 b. 4 × R = 300; R = 75

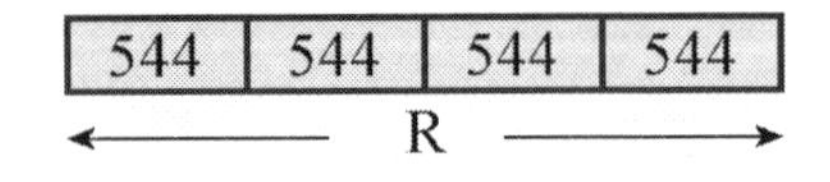

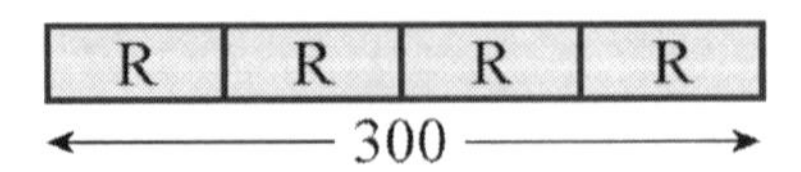

10.

Divisible by	2	3	5	6	9
534	X	X		X	
123		X			

Divisible by	2	3	5	6	9
1,605		X	X		
2,999					

11. a. 21 = 3 × 7 b. 12 = 2 × 2 × 3 c. 38 = 2 × 19
    d. 75 = 3 × 5 × 5 e. 124 = 2 × 2 × 31 f. 89 = 1 × 89

Chapter 2: Large Numbers and the Calculator

A Little Bit of Millions, p. 75

1.

a. 724,000,000 724 million	b. 512,000,000 512 million	c. 404,000,000 404 million
d. 4.000,000 4 million	e. 86,000,000 86 million	f. 8,345,000 8 million 345 thousand
g. 22,906,000 22 million 906 thousand	h. 514,310,000 514 million 310 thousand	i. 40,300,000 40 million 300 thousand

2. a. 18 million

	1	8,	0	0	0 ,	0	0	0

b. 906 million

9	0	6,	0	0	0 ,	0	0	0

c. 2 million 400 thousand

		2,	4	0	0,	0	0	0

d. 70 million 90 thousand

	7	0,	0	9	0,	0	0	0

3.

a. 779,453,230 779 million 453 thousand 230	b. 9,290,807 9 million 290 thousand 807
c. 52,907,033 52 million 907 thousand 33	d. 55,001,453 55 million 1 thousand 453
e. 72,025,090 72 million 25 thousand 90	f. 228,010,200 228 million 10 thousand 200

g. 107,000,453 107 million ____ ~~thousand~~ 453	h. 72,000,090 72 million ____ ~~thousand~~ 90
i. 28,000,006 28 million ____ ~~thousand~~ 6	j. 370,000,018 370 million ____ ~~thousand~~ 18

4. a. 41,456,200 b. 80,080,080 c. 5,006,170 d. 299,003,009

5.

a.	b.	c.	d.
930,000	500,000	999,994	999,600
940,000	600,000	999,995	999,700
950,000	700,000	999,996	999,800
960,000	800,000	999,997	999,900
970,000	900,000	999,998	1,000,000
980,000	1,000,000	999,999	
990,000		1,000,000	
1,000,000			

6.

a. 6,111,050 > 5,990,099	b. 2,223,020 > 2,222,322	c. 192,130,659 < 192,130,961
d. 18,000,0000 > 181,000	e. 13,395,090 < 13,539,099	f. 2,367,496 > 988,482
g. 6,009,056 < 6,090,045	h. 1,000,999 < 1,001,000	i. 17,199,066 < 71,857,102

A Little Bit of Millions, cont.

7. Answers will vary. Please check the students' work.

8. Answers will vary. Please check the students' work.

Puzzle corner warm-up questions:
100 goes into 1,000 ten times. 1,000 goes into 10,000 ten times.
100 goes into 10,000, one hundred times.
10,000 goes into 100,000 ten times. 100,000 goes into 1,000,000 ten times.
10,000 goes into 1,000,000 one hundred times.

Puzzle corner: Yes, Jack was correct. Yes.

Place Value Up to Billions, p. 78

1. "eighty-five billion, three hundred fifty-nine million, two hundred four thousand, thirty-one."
   a. 2 b. 8 c.9 d. 5

2. a. 39,204,848,486
   "thirty-nine billion, two hundred four million, eight hundred forty-eight thousand, four hundred eighty-six."
   b. 490,255,549,632
   "four hundred ninety billion, two hundred fifty-five million, five hundred forty-nine thousand, six hundred thirty-two."
   c. 2,843,729,584
   "two billion, eight hundred forty-three million, seven hundred twenty-nine thousand, five hundred eighty-four."
   d. 309,082,048,392
   "three hundred nine billion, eighty-two million, forty-eight thousand, three hundred ninety-two."

3. a. 308,067,008,307 = 308 billion, 67 million, 8 thousand, 307
   b. 45,038,300,820 = 45 billion, 38 million, 300 thousand, 820
   c. 915,008,360,000 = 915 billion, 8 million, 360 thousand
   d. 9,000,004,000 = 9 billion, 4 thousand

4. a. 159,372,932,002 b. 7,372,040,020 c. 372,000,150 d. 607,000,043,017

5. a. 32,030,200 b. 500,000,500,005 c. 612,087,002,300 d. 45,000,003,043

6. a. 302,478 = 3 × 100,000 + 0 × 10,000 + 2 × 1,000 + 4 × 100 + 7 × 10 + 8 × 1
   b. 269,115 = 2 × 100,000 + 6 × 10,000 + 9 × 1,000 + 1 × 100 + 1 × 10 + 5 × 1
   c. 6,087,240 = 6 × 1,000,000 + 0 × 100,000 + 8 × 10,000 + 7 × 1,000 + 2 × 100 + 4 × 10 + 0 × 1
   d. 87,034 = ☐ × 1,000,000 + ☐ × 100,000 + 8 × 10,000 + 7 × 1,000 + 0 × 100 + 3 × 10 + 4 × 1
   e. 2,167,900 = 2 × 1,000,000 + 1 × 100,000 + 6 × 10,000 + 7 × 1,000 + 9 × 100 + 0 × 10 + 0 × 1
   f. 97,225 = 9 × 10,000 + 7 × 1,000 + 2 × 100 + 2 × 10 + 5 × 1
   g. 708,340 = 7 × 100,000 + 0 × 10,000 + 8 × 1,000 + 3 × 100 + 4 × 10 + 0 × 1

7.

a. 293,476,020 Place: *ten thousands place* Value: 70,000 (seventy thousand)	b. 3,299,005,392 Place: millions place Value: 9,000,000 (nine million)
c. 28,837,402,000 Place: ten billions place Value: 20,000,000,000 (twenty billion)	d. 293,476,020 Place: ten millions place Value: 90,000,000 (ninety million)
e. 3,299,005,392 Place: hundred millions place Value: 200,000,000 (two hundred million)	f. 28,837,432,000 Place: ten thousands place Value: 30,000 (thirty thousand)

8. a. 801,041,000,000 b. 6,907,057,000 c. 245,048,035

Exponents and Powers, p. 81

1.

a. $5^2 = 5 \times 5 = 25$ b. $2^3 = 2 \times 2 \times 2 = 8$ c. $3^3 = 3 \times 3 \times 3 = 27$ d. $10^2 = 10 \times 10 = 100$	e. $10^3 = 10 \times 10 \times 10 = 1{,}000$ f. $7^2 = 7 \times 7 = 49$ g. $2^4 = 2 \times 2 \times 2 \times 2 = 16$ h. $1^6 = 1 \times 1 \times 1 \times 1 \times 1 \times 1 = 1$

2.

a. $4 \times 4 \times 4 = 4^3 = 64$ b. $9 \times 9 = 9^2 = 81$ c. $10 \times 10 \times 10 \times 10 = 10^4 = 10{,}000$ d. five to the third power = $5^3 = 125$	e. $1 \times 1 \times 1 \times 1 \times 1 = 1^5 = 1$ f. $2 \times 2 \times 2 \times 2 \times 2 = 2^5 = 32$ g. $3 \times 3 \times 3 \times 3 = 3^4 = 81$ h. zero to the tenth power = $0^{10} = 0$

3.

a. $2 + 2 + 2 + 2 = 4 \times 2 = 8$ $2 \times 2 \times 2 \times 2 = 2^4 = 16$	b. $5 + 5 + 5 = 3 \times 5 = 15$ $5 \times 5 \times 5 = 5^3 = 125$

4.

a. $10^2 = 100$ b. $10^3 = 1{,}000$ c. $10^4 = 10{,}000$ d. $10^5 = 100{,}000$	e. $10^6 = 1{,}000{,}000$ f. $10^7 = 10{,}000{,}000$ g. $10^8 = 100{,}000{,}000$ h. $10^9 = 1{,}000{,}000{,}000$

SHORTCUT: In any power of ten, such as 10^8, the exponent tells us how many zeros the number has after the digit 1.

5. a. 6,000,000 b. 16,000,000 c. 1,500,000,000
d. 1,500,000 e. 27,000,000 f. 60,000,000
g. 24,000,000 h. 80,000,000 i. 8,000,000,000

6.

a. $5 \times 10^2 = 500$ $5 \times 10^3 = 5{,}000$ $5 \times 10^4 = 50{,}000$	b. $7 \times 10^6 = 7{,}000{,}000$ $2 \times 10^4 = 20{,}000$ $6 \times 10^7 = 60{,}000{,}000$	c. $51 \times 10^3 = 51{,}000$ $161 \times 10^6 = 161{,}000{,}000$ $29 \times 10^4 = 2{,}90{,}000$

7.

a. $200 \times 3{,}000$ is equal to $2 \times 100 \times 3 \times 1{,}000$, which is equal to $2 \times 3 \times 100 \times 1{,}000$ $= 6 \times 100{,}000 = 600{,}000$	b. $6{,}000 \times 200 \times 50$ is equal to $(6 \times 1000) \times (2 \times 100) \times (5 \times 10)$ $= 6 \times 2 \times 5 \times 1000 \times 100 \times 10$ $= 60 \times 1{,}000{,}000$ $= 60{,}000{,}000$

8.

a. $6 \times 10^3 = 6{,}000$ $71 \times 10^6 = 71{,}000{,}000$	b. $3 \times 10^5 = 300{,}000$ $9 \times 10^7 = 90{,}000{,}000$	c. $56 \times 10^4 = 560{,}000$ $295 \times 10^7 = 2{,}950{,}000{,}000$

Exponents and Powers, cont.

9. Shortcut: Add the exponents to find how many zeros to tag onto the answer.

a. $10^3 \times 10^2 = 100{,}000$	b. $5 \times 10^2 \times 10^4 = 5{,}000{,}000$
c. $10^5 \times 10^3 = 100{,}000{,}000$	d. $8 \times 10^4 \times 2 \times 10^3 = 160{,}000{,}000$
e. $10^6 \times 10^2 \times 10^2 = 10^{10}$	f. $10^3 \times 10^5 \times 10^2 \times 10^4 = 10^{14}$

10. Pluto's surface area is about 17,000,000 km^2.
The Sun's average distance from Earth is 150,000,000 km.

Haumea is a dwarf planet located beyond Neptune's orbit.
The mass of Haumea is about 4,000,000,000,000,000,000,000 kg.

Puzzle corner: The Sun is about 1,000 times heavier than Jupiter.

Adding and Subtracting Large Numbers, p. 84

1.

	a. 90,000	b. 99,000,000	c. 999,000
+ 1,000	91,000	99,001,000	1,000,000
+ 10,000	100,000	99,010,000	1,009,000
+ 100,000	190,000	99,100,000	1,099,000
+ 1,000,000	1,090,000	100,000,000	1,999,000

2.

a.	b.
1/2 million - 500,000 a hundred hundreds - 10^4 1/10 million - 100,000 1/4 million - 250,000 3/4 million - 750,000 a thousand thousands - 10^6 2/10 million - 200,000	1 million − 50,000 = 950,000 1 million − 500,000 = 500,000 10^8 - 100,000,000 1 billion − 500 million = 1/2 billion 1 billion − 50 million = 950,000,000 1 million − 5,000 = 995,000 1 billion − 5 million = 995,000,000

3. a. 3,138,270,093 b. 50,878,749 c. 92,915,820 d. 254,349,802
e. 456,580,000 f. 843,010 g. 17,868,810 h. 3,910,729,609

4.

a. 1 million − 100 thousand = 900,000 1 million − 10 thousand = 990,000 1 million − 1 thousand = 999,000	b. 7 million − 500 thousand = 6,500,000 7 million − 50 thousand = 6,950,000 7 million − 5 thousand = 6,995,000

Adding and Subtracting Large Numbers, cont.

5.

a.	b.	c.
458,000,000 468,000,000 478,000,000 488,000,000 498,000,000 508,000,000 518,000,000 528,000,000 538,000,000 548,000,000	79,650,000 79,800,000 79,950,000 80,100,000 80,250,000 80,400,000 80,550,000 80,700,000 80,850,000 81,000,000	450,996,000 450,997,000 450,998,000 450,999,000 451,000,000 451,001,000 451,002,000 451,003,000 451,004,000 451,005,000
Each difference is 10,000,000	Each difference is 150,000	Each difference is 1,000

6.

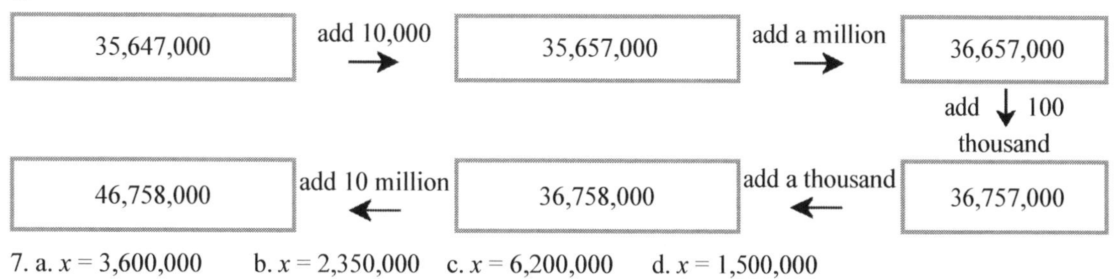

7. a. $x = 3{,}600{,}000$ b. $x = 2{,}350{,}000$ c. $x = 6{,}200{,}000$ d. $x = 1{,}500{,}000$

Rounding, p. 87

1.

number	2,017,249	38,802,155	82,009,709	217,299,204
to the nearest 1,000	2,017,000	38,802,000	82,010,000	217,299,000
to the nearest 10,000	2,020,000	38,800,000	82,010,000	217,300,000
to the nearest 100,000	2,000,000	38,800,000	82,000,000	217,300,000
to the nearest million	2,000,000	39,000,000	82,000,000	217,000,000

2.

number	24,302	496,253	299,389,932	2,505,899,430
to the nearest 1,000	24,000	496,000	299,390,000	2,505,899,000
to the nearest 10,000	20,000	500,000	299,390,000	2,505,900,000
to the nearest 100,000	0	500,000	299,400,000	2,505,900,000
to the nearest million	0	0	299,000,000	2,506,000,000

3. a. Of the seven dots, the fifth and sixth are rounded to 57,000.
   b. The range of numbers is from 56,500 to 57,499. It is also shown on the number line below:

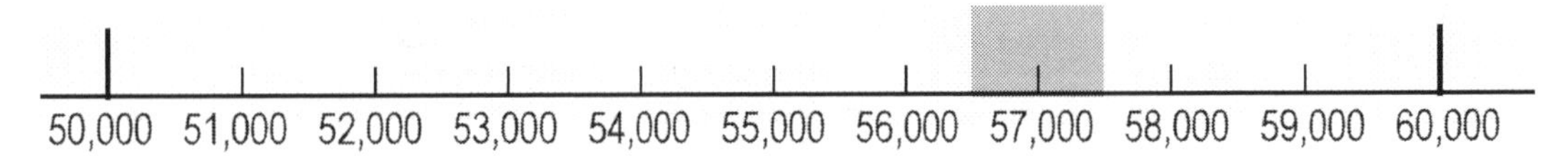

4. The range of numbers is from 125,000 to 134,999.

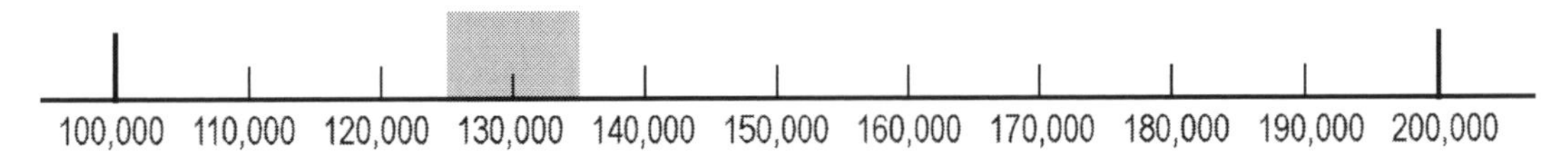

5. The range of numbers is from 350,000 to 449,999.

6.

a. The two towns have approximately 470,000 people in all. There are about 50,000 more people in Buffalo than in Rochester.
b. New York had about 16,000,000 more people than Utah.
c. Rounding to the nearest ten thousand makes the problem simple. Annie's yearly income was \$61,224. ≈ \$60,000. So, she earns about \$5,000 monthly.
d. The simplest way is to round the annual mileage to the nearest ten. Mike's internet service bill is \$78.84 ≈ \$80.00 each month. He pays about \$960 for the internet in a year.

7. a.

MARITAL STATUS	
Never married	60,000,000
Now married (not separated)	120,000,000
Separated	5,000,000
Widowed	15,000,000
Divorced	22,000,000

Source: From Census 2000 data, www.census.gov.

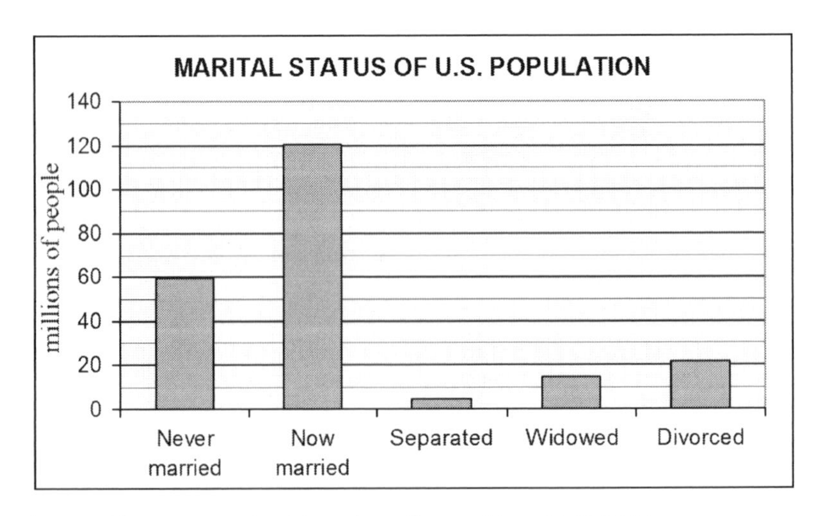

b. The *estimated* number of people who are either separated, widowed, or divorced is ≈ 42,000,000.
c. The *estimated* number of people who are not married is ≈ 102,000,000.

The Calculator and Estimating, p. 90

1.

a. 54,395 + 89,302 (round to thousands) My estimation: 54,000 + 89,000 = 143,000 Exact answer: 143,697 Error of estimation: 697	b. 9,807,520 − 1,532,392 (round to millions) My estimation: 10,000,000 − 2,000,000 = 8,000,000 Exact answer: 8,275,128 Error of estimation: 275,128
c. 1,224,845 (to millions) ÷ 995 (to thousands) My estimation: 1,000,000 ÷ 1,000 = 1,000 Exact answer: 1,231 Error of estimation: 231	d. 2,873 × 3,204 (round to thousands) My estimation: 3,000 × 3,000 = 9,000,000 Exact answer: 9,205,092 Error of estimation: 205,092
e. 2,793 × 423 My estimation: 3,000 × 400 = 1,200,000 Exact answer: 1,181,439 Error of estimation:18,561	f. 132 × 49 × 8,231 My estimation: 100 × 50 × 8,000 = 40,000,000 Exact answer: 53,238,108 Error of estimation: 13,238,108

2.

a. My estimation: 16 × $50 = $800 Exact answer: $744 Error of estimation: $56	b. My estimation: 6 × $1.50 + $10 = $19 Exact answer: $19.10 Error of estimation: $0.10

3.

a. $5^1 = 5$ $5^2 = 5 \times 5 = 25$ $5^3 = 5 \times 5 \times 5 = 125$ $5^4 = 5 \times 5 \times 5 \times 5 =$ 625	b. $5^5 =$ 3,125 $5^6 =$ 15,625 $5^7 =$ 78,125 $5^8 =$ 390,625	c. $5^9 =$ 1,953,125 $5^{10} =$ 9,765,625 $5^{11} =$ 48,828,125 $5^{12} =$ 244,140,625

4. $5^9 =$ 1,953,125

5.

a. $8^1 = 8$ $8^2 = 8 \times 8 = 64$ $8^3 = 8 \times 8 \times 8 =$ 512 $8^4 =$ 4,096	b. $8^5 =$ 32,768 $8^6 =$ 262,144 $8^7 =$ 2,097,152 $8^8 =$ 16,777,216	c. $8^9 =$ 134,217,728 $8^{10} =$ 1,073,741,824 $8^{11} =$ 8,589,934,592 $8^{12} =$ 68,719,476,736

6. a. $8^7 =$ 2,097,152 b. $8^{10} =$ 1,073,741,824

7.

a. $12^{6} > 1,000,000$	b. $8^{7} > 1,000,000$	c. $42^{4} > 1,000,000$
d. $11^{6} > 1,000,000$	e. $2^{20} > 1,000,000$	f. $0^{\square} > 1,000,000$ **impossible**
g. $10^{6} = 1,000,000$	h. $1^{\square} > 1,000,000$ **impossible**	i. $100^{3} = 1,000,000$

8. a. Shortcut for repeated addition is multiplication. 20 + 20 + 20 + 20 + 20 + 20 is 6 × 20.
   b. You would need to add 20 50,000 times in order to reach 1 million.
   c. You would need to add 40 25,000 times in order to reach 1 million.
   d. You would need to add 5,000 200 times in order to reach 1 million.

9. a. (× 2) 1 2 4 8 16 **32 64 128 256 512**
   b. (× 2) 6 12 24 48 96 **192 384 768 1,536 3,072**
   c. (× 6) 12 72 432 2,592 **15,552 93,312 559,872 3,359,232 20,155,392**
   d. (× 3) 18 54 162 486 **1,458 4,374 13,122 39,366 118,098**
   e. (× 11) 2 22 242 2,662 **29,282 322,102 3,543,122 38,974,342 428,717,762**
   f. (÷ 10) 1,000,000,000 100,000,000 10,000,000 1,000,000 **100,000 10,000 1,000 100 10**
   g. (÷ 2) 1,048,576 524,288 262,144 131,072 **65,536 32,768 16,384 8,192 4,096**

When to Use the Calculator p. 93

1. a. about 1,000 km a day; mental math b. about 500 km; mental math c. $4,375.00; calculator
   d. 110,800 km; mental math e. about 7 hours; calculator f. about 2,700 hours; calculator

2. a. About 9,660 passengers per day b. About 420 passengers per bus per day

3. a. About 468 gallons used per person b. About 16,018,275 gallons of fuel was used per hour

4. a. $9,500 / year × 15 years = $142,500.00 b. $9,500 / year ÷ 365.25 days/year ≈ $26.00

Mixed Review, p. 95

1.

a. 76 − 65 = 11 subtrahend	b. 57 − 39 = 18 minuend	c. 48 − 29 = 19 difference

2. $x + 9{,}380 + 3{,}928 = 93{,}450$; $x = 80{,}142$

3. a. 84,000 b. 132,000,000 c. 300,000,000 d. 10,000 e. 27 f. 7,000,000

4. 10^6

5. There are 8,760 hours in a year.
   Estimate: 360 × 30 = 10,800

6. a. 128; 43 × 128 = 5,504
   b. 95; 82 × 95 = 7,790

7. a. 2 × 2 × 7 b. 2 × 7 × 7 c. 2 × 3 × 11 d. 1 × 17 e. 3 × 17 f. 1 × 53

Review, p. 97

1. a. 9,070,560 b. 60,007,540 c. 50,000,050,050 d. 431,098,000,940

2.

a. 405,229,020 Place: ten thousands Value: twenty thousand	b. 97,024,003,245 Place: one millions Value: four million
c. 230,560,079,000 Place: ten billions Value: thirty billion	d. 4,589,211,000 Place: hundred million Value: five hundred million

3.

number	69,066	14,506,439	389,970,453	12,976,895,322
to the nearest 1,000	69,000	14,506,000	389,970,000	12,976,895,000
to the nearest 10,000	70,000	14,510,000	389,970,000	12,976,900,000
to the nearest 100,000	100,000	14,500,000	390,000,000	12,976,900,000
to the nearest million	0	15,000,000	390,000,000	12,977,000,000

4.

a. $8^2 = 8 \times 8 = 64$	d. $1^5 = 1 \times 1 \times 1 \times 1 \times 1 = 1$
b. $4^3 = 4 \times 4 \times 4 = 64$	e. $100^2 = 100 \times 100 = 10{,}000$
c. $10^3 = 10 \times 10 \times 10 = 1{,}000$	f. $2^5 = 2 \times 2 \times 2 \times 2 \times 2 = 32$

5.

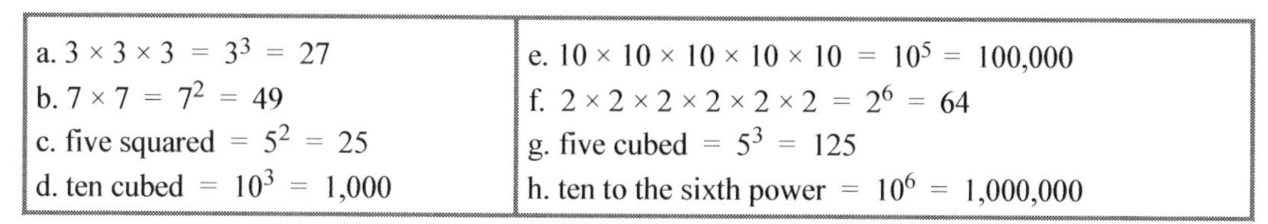

a. $3 \times 3 \times 3 = 3^3 = 27$	e. $10 \times 10 \times 10 \times 10 \times 10 = 10^5 = 100{,}000$
b. $7 \times 7 = 7^2 = 49$	f. $2 \times 2 \times 2 \times 2 \times 2 \times 2 = 2^6 = 64$
c. five squared $= 5^2 = 25$	g. five cubed $= 5^3 = 125$
d. ten cubed $= 10^3 = 1{,}000$	h. ten to the sixth power $= 10^6 = 1{,}000{,}000$

6. a. 22,000,000 b. 3,600,000 c. 10,000,000,000 d. 6,000
e. 800,000 f. 200,000,000 g. 210,000 h. 829,000,000

7. 8,079,083

8.

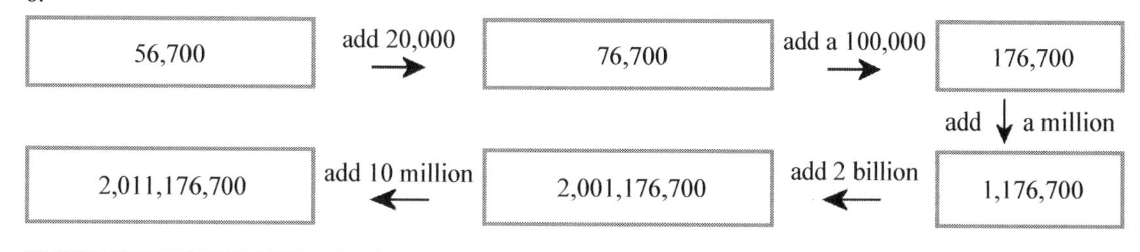

9. a. $6{,}808 + 493{,}420$ My estimation: 7,000 + 500,00 = 507,000 Exact answer: 500,228 Error of estimation: 6,772	b. $3{,}703 \times 52{,}380$ My estimation: $4{,}000 \times 50{,}000$ = 200,000,000 Exact answer: 193,963,140 Error of estimation: 6,036,860

10. \$8,881,833,600 weekly

11. \$36 billion or \$36,000,000,000 (Multiply \$9,000 and 4,000,000.)

Mystery Number: a. 2,260,430 b. 3,023,183

Chapter 3: Problem Solving

Balance Problems and Equations, p. 102

1.

a. Equation: $9 = \square + 3$ Solution: $\square = 6$	b. Equation: $3\bigcirc = 21$ Solution: $\bigcirc = 7$
c. Equation: $2\square + 2 = 16$ Solution: $\square = 7$	d. Equation: $\square + 7 = 51$ Solution: $\square = 44$

2.

a. Equation: $24 + 7 = x$ Solution: $x = 31$	b. Equation: $38 = x + 12$ Solution: $x = 26$
c. Equation: $88 = 2x$ Solution: $x = 44$	d. Equation: $16 + 6 = 2x$ Solution: $x = 11$
e. Equation: $3x = 6 + 32 + 4$ Solution: $x = 14$	f. Equation: $2x + 6 = 32 + 4$ Solution: $x = 15$

3.

a. $2x + 7 = x + 19$ $x + 7 = 19$ $x = 12$	b. $2x + 47 = 3x$ $47 = x$ $x = 47$
c. $2x = x + 17$ $x = 17$	d. $3x + 7 = 2x + 23$ $x + 7 = 23$ $x = 16$
e. $2x + 44 = 4x$ $44 = 2x$ $x = 22$	f. $3x + 8 = x + 14$ $2x + 8 = 14$ $x = 3$

4.

a. $2x + 3 = 93$ | next, remove 3 from both sides $2x = 90$ $x = 45$	b. $x + 51 = 2x + 5$ | next, remove x from both sides $51 = 5 + x$ $x = 46$
c. $3x + 9 = 27$ | next, remove 9 from both sides $3x = 18$ $x = 6$	d. $3x + 9 = x + 27$ | next, remove x from both sides $2x = 18$ $x = 9$
e. $x + 9 + 6 = 2x + 2$ | next, remove x from both sides $9 + 6 = x + 2$ $15 = x + 2$ $x = 13$	f. $4x + 6 = x + 13 + 5$ | next, remove x from both sides $3x = 12$ $x = 4$

Balance Problems and Equations, cont.

5.

a. $2x + 5 = 41$ | next, remove 5 from both sides $2x = 36$ $x = 18$	b. $3x + 37 = 4x$ | next, remove $3x$ from both sides $37 = x$ $x = 37$
c. $x + 15 = 2x + 7$ | next, remove x from both sides $15 = x + 7$ $x = 8$	d. $3x + 8 = 26$ | next, remove 8 from both sides $3x = 18$ $x = 6$

More Equations, p. 107

1. a. $6y + 12 = 78$; $y = 11$ b. $6 \times 12 + y = 78$; $y = 6$

2. a. $4x + 142 = 298$; $x = 39$ b. $2x + 120 = 230$; $x = 55$

3.

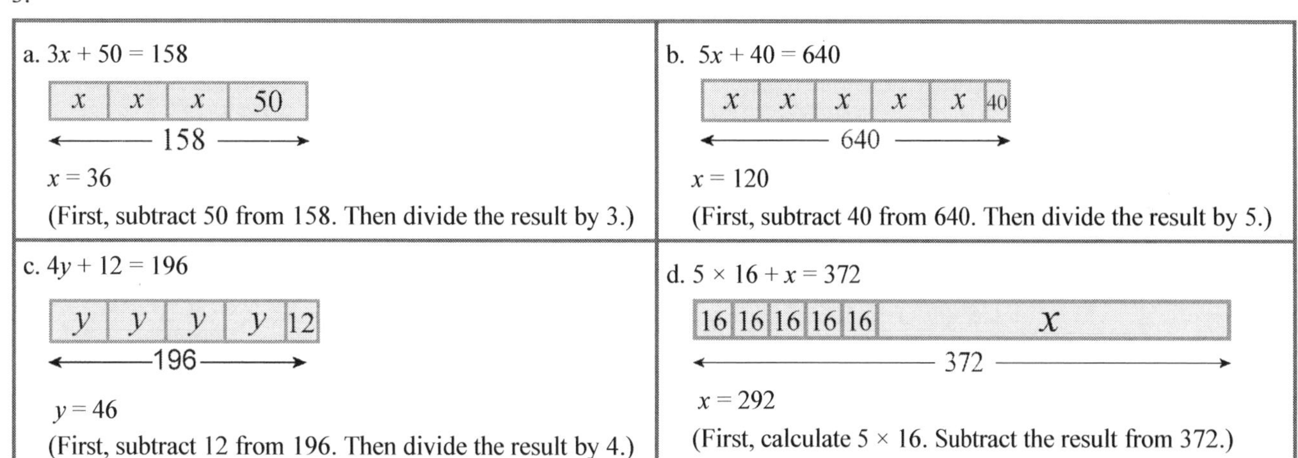

a. $3x + 50 = 158$ x x x 50 ← 158 → $x = 36$ (First, subtract 50 from 158. Then divide the result by 3.)	b. $5x + 40 = 640$ x x x x x 40 ← 640 → $x = 120$ (First, subtract 40 from 640. Then divide the result by 5.)
c. $4y + 12 = 196$ y y y y 12 ← 196 → $y = 46$ (First, subtract 12 from 196. Then divide the result by 4.)	d. $5 \times 16 + x = 372$ 16 16 16 16 16 x ← 372 → $x = 292$ (First, calculate 5×16. Subtract the result from 372.)

4. a. $4x + 4 = 28$; $x = \$6$ b. $6x + 18 = 36$; $x = \$3$
   c. $3x + 7 \times 5 = 80$; $x = \$15$ d. $7 \times 55 + x = 265$; $x = 45$ cm

5. a. 103 b. 242 c. 89 d. 105 e. 1,071 f. 506

6. a. Its perimeter is 9 feet. b. It is 11 feet 4 inches.
   c. 1,156 square inches

7.

a. $8 \times Z = 536$ $Z \times 8 = 536$ $536 \div 8 = Z$ $536 \div Z = 8$	b. $29 \times 15 = U$ $15 \times 29 = U$ $U \div 15 = 29$ $U \div 29 = 15$	c. $W \times 3 = 732$ $3 \times W = 732$ $732 \div W = 3$ $732 \div 3 = W$

8. a. $Z = 67$ b. $U = 435$ c. $W = 244$

Problem Solving with Bar Models 1, p. 111

1. $125 ÷ 5 = $25; $25 × 4 = $100. The the new price is $100.

2. 680 g ÷ 5 = 136 g; 136 g × 2 = 272 g. Two pieces weigh 272 g.

3. They cost $5.00 First, solve the price of one bottle of juice:
   $1.50 ÷ 3 = $0.50; 2 × $0.50 = $1.00.
   Two bottles of water and two juices cost 2 × $1.00 + 2 × $1.50 = $5.00

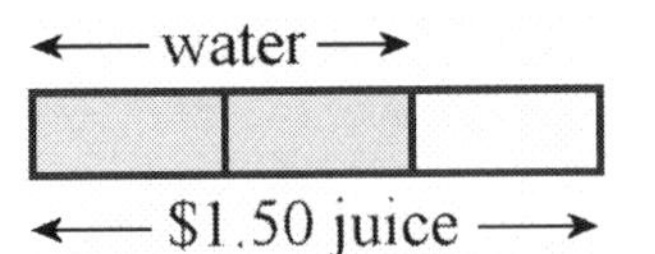

4. $12 ÷ 6 = $2. $12 + $2 = $14. The new price is $14.

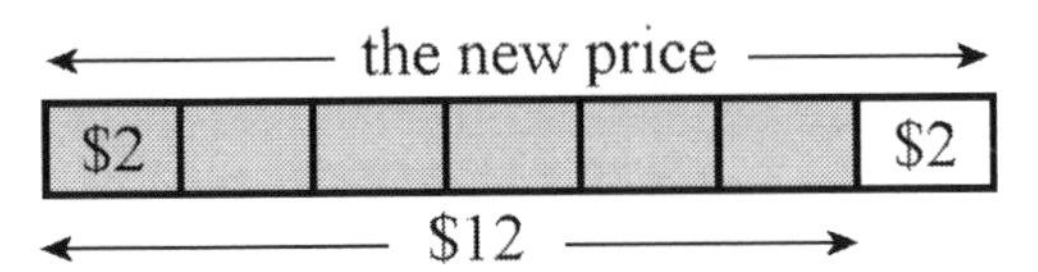

5. $4.50 ÷ 5 = $0.90. $4.50 + $0.90 = $5.40. The new price is $5.40.

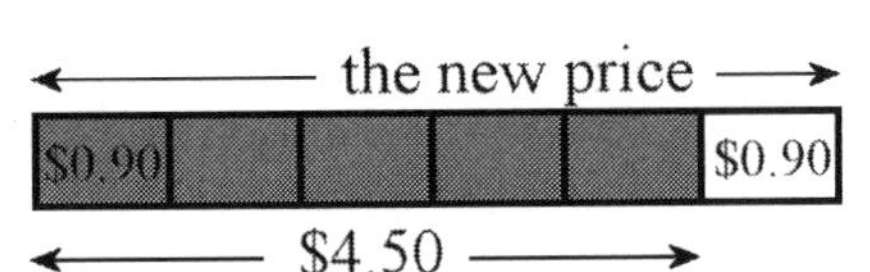

6. a. $1.00 ÷ 10 = $0.10. $1.00 + $0.10 = $1.10. It will cost $1.10.
   b. A round-trip in May cost $2, and in June $2.20 It will cost $0.20 more in June.

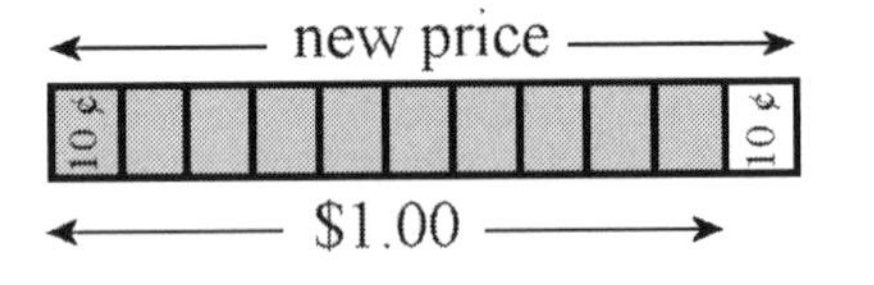

7. $10.50 ÷ 5 = $2.10. 3 × $2.10 = $6.30. The discounted price for one shirt is $6.30. Annie's total bill is 10 × $6.30 = $63.00.

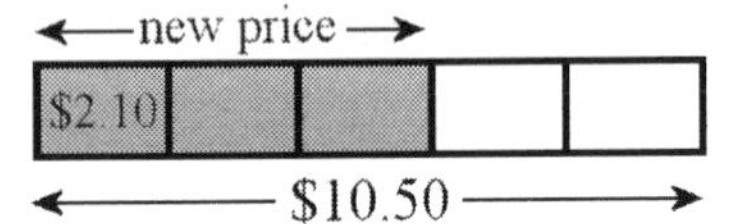

8. One block in the model is 3,687 ÷ 3 = 1,299. The population of Eagleby is 5 × 1,299 = 6,145. The population of both towns is 6,145 + 3,687 = 9,832.

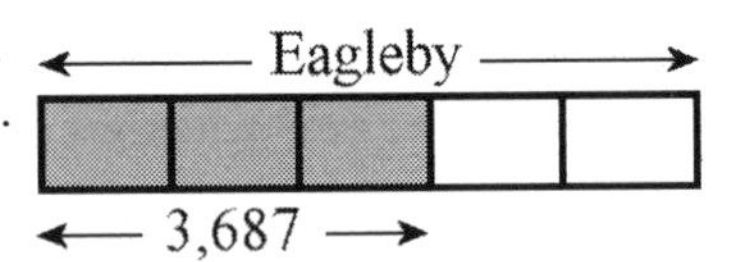

9. One-fifth of the price of the package of the small envelopes is $2.50 ÷ 5 = $0.50. Two-fifths is then double that, or $1.00.
   The package of 10 large ones costs $2.50 + $1.00 = $3.50.
   For 50 envelopes of each kind, we need five of each package.
   The total cost is then 5 × $2.50 + 5 × $3.50 = $30.

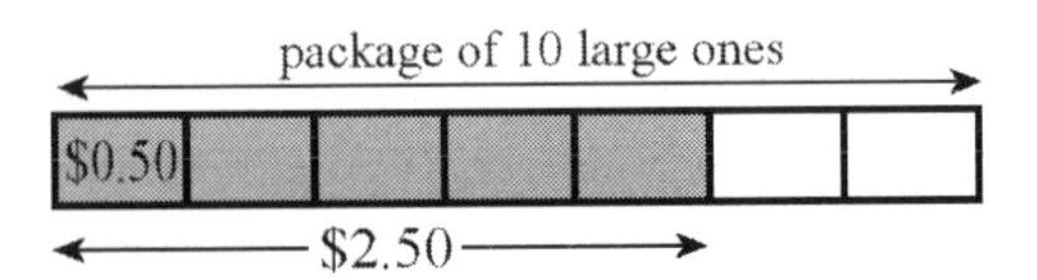

Problem Solving with Bar Models 2, p. 114

1. The two parts in the model that are the same size are 420 − 18 = 402 in total.
   The box with fewer apples has 201 apples.

2. The two parts in the model that are the same size are
   \$590 − \$100 = \$490 in total. One of those parts is \$490 ÷ 2 = \$245.
   The more expensive saw cost \$345 and the cheaper saw cost \$245.

3. The two parts in the model that are the same size are
   1,689 − 155 = 1,534 in total. One of those parts is 1,534 ÷ 2 = 767.
   There were 767 visitors in June, and 922 visitors in July.

4. The two parts in the model that are the same size are 210 − 30 = 180 in total.
   One of those parts would be 180 ÷ 2 = 90.
   The shorter piece of string is 90 cm long, and the longer piece is 120 cm long.
   The shorter piece was divided into two equal pieces, which are
   90 cm ÷ 2 = 45 cm long. The longer piece was divided into three equal pieces,
   which are 120 cm ÷ 3 = 40 cm long.

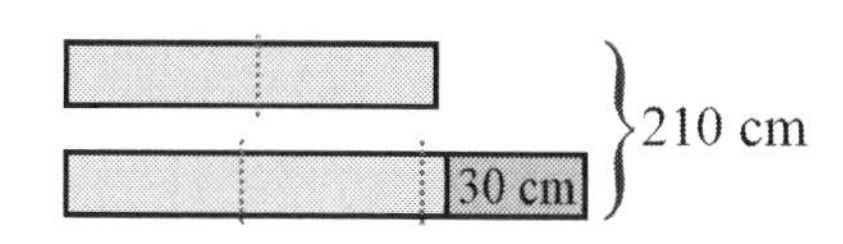

5. The three parts in the model that are the same size are 119 − 20 − 3 = 96
   in total. One of them is \$96 ÷ 3 = \$32. Carla has \$32.
   Beth has \$32 + \$20 = \$52. Amy has \$32 + \$3 = \$35.

Puzzle Corner:
If Sandy had earned \$25 more than she did, all three would have earned the same amount, and the total would have been \$285. So, dividing \$285 ÷ 3 = \$95, we get what Allison and Peter earned. Sandy earned \$25 less than that, or \$70.

Problem Solving with Bar Models 3, p. 116

Teaching box:
To solve this, divide the total length of the fence by four. That gives you the length of the fence Alice painted. So, Alice painted 31 ft of the fence, and Hannah painted 93 ft.

1. There are three parts. One part is \$9,000 ÷ 3 = \$3,000. The elder received \$6,000
   and the younger received \$3,000.

2. The book's weight is 9 parts, and the packaging is 1 part.
   So, there are a total of 10 parts. One part is 2,200 g ÷ 10
   = 220 g. The book weighs 9 × 220 g = 1,980 g.

3. The price of the energy-saving bulb is 3 parts and the price of the cheap
   one is one part in the bar model. So, there are four parts in total. Divide
   \$8.40 ÷ 4 = \$2.10 to find one part. The energy-saving bulb costs
   3 × \$2.10 = \$6.30. Five of them cost 5 × \$6.30 = \$31.50.

Problem Solving with Bar Models 3, cont.

4. Miles traveled by an airplane flying with a constant speed

Hours	Miles
1	550
2	1,100
3	1,650
4	2,200
5	2,750

Price of chairs

Chairs	Price
1	$18
2	$36
3	$54
4	$72
5	$90

Weight of identical boxes

Boxes	Weight
10	130 kg
20	260 kg
30	390 kg
40	520 kg
50	650 kg

5. $1,355 ÷ 5 = $271. 2 × $271 = $542. Two paintings cost $542.

6. The price of one can is $2.50 ÷ 2 = $1.25. Five cans cost 5 × $1.25 = $6.25.

7. a. We will find 3/4 of the side lengths. First, 3/4 of 16 cm: 16 cm ÷ 4 × 3 = 12 cm. Then, 3/4 of 40 cm: 40 cm ÷ 4 × 3 = 30 cm. The perimeter of the smaller rectangle is 2 × (12 cm + 30 cm) = 84 cm. The perimeter of the original rectangle is 2 × (16 cm + 40 cm) = 112 cm.

   b. Yes. Calculate 3/4 of 112 cm to verify this: 112 cm ÷ 4 × 3 = 84 cm. Of course, that gives a quicker way to calculate the answer to part (a).

Problem Solving with Bar Models 4, p. 118

1. One block is 840 lb ÷ 2 = 420 lb. Jake's car weighs 3 × 420 lb = 1,260 lb.

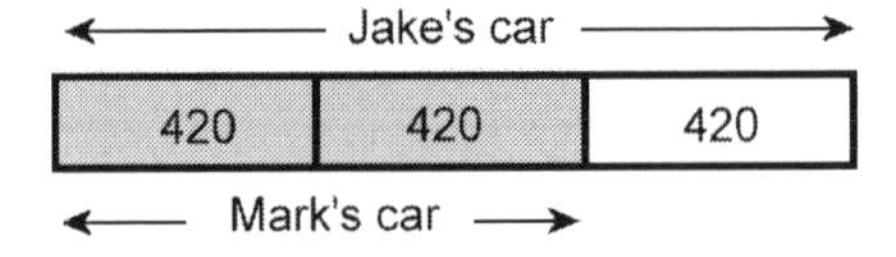

2. One block is (6 ft 6 in.) ÷ 6 = 1 ft 1 in. Henry is 5 × (1 ft 1 in.) = 5 ft 5 in. tall.

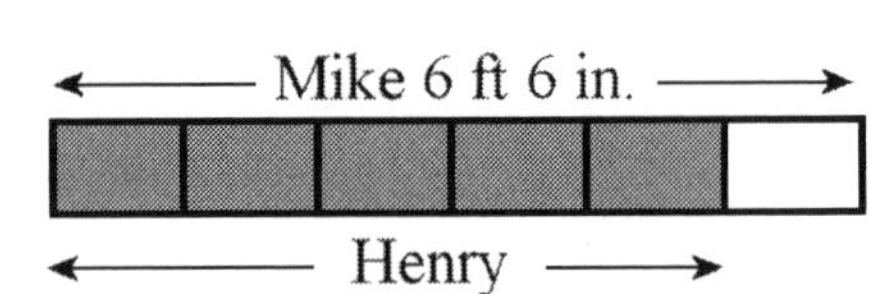

3. To find one block in the bar model, we divide 120 cm ÷ 5 = 24 cm. Then, Ryan's dad is 8 × 24 cm = 192 cm tall.

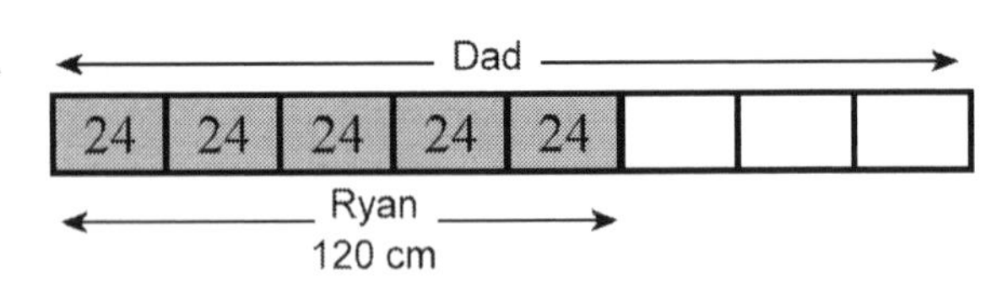

4. The bouquet costs $16.50. First, we find the price of one rose, so we find 5/6 of $1.80: $1.80 ÷ 6 × 5 = $1.50. The bouquet of 5 roses and 5 lilies costs 5 × $1.50 + 5 × $1.80 = $16.50.

5. The girls spent $22.50 together. Emily spent 4 × 4.50 = $18. Together, the girls spent $18 + $4.50 = $22.50.

6. The second shirt costs $11.50 − $2.55 = $8.95. The third shirt costs $8.95 − $2 = $6.95. The cost of all three is $11.50 + $8.95 + $6.95 = $29.95.

7. His original paycheck was ($250 + $660) × 2 = $1,820.

← 1/2 → ← 1/2 →
$660 | $250 |
← paycheck →

8. Three-eighths of 40 cards is 40 ÷ 8 × 3 = 15. So, Tyler now has 40 − 15 = 25 cards and Chris has 10 + 15 = 25.

9. The new price for one book is ($162 + $36) ÷ 18 = $11.

10. The side of his original square is 264 ÷ 4 = 66 pixels. The side of his new square is 66 + 15 = 81 pixels. The perimeter of the new square is 4 × 81 = 324 pixels.

Puzzle Corner: The discounted price (for a 6-month subscription) is 3/4 of the original. We can find 1/4 of the original price by dividing the discounted price by 3: $54 ÷ 3 = $18. Now, the original price is four times that, or $72. Then, one month would have cost 1/6 of that, or $12.

Mixed Review, p. 122

1. $x = 547 - 119 - 38 = 390$. See the bar model on the right.

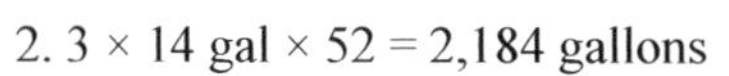

2. 3 × 14 gal × 52 = 2,184 gallons

3. a. 16 − 7 = 9 b. $3 + 9 + y = 20$

4. a. 14 b. 10 c. 71

5. c. 7 × 65 + 3

6. a. the same b. not the same c. not the same

7. a. 64 = 2 × 2 × 2 × 2 × 2 × 2
   b. 60 = 2 × 2 × 3 × 5
   c. 85 = 5 × 17

8.

2 × 79 = 158 3 × 79 = 237 4 × 79 = 316 5 × 79 = 395 6 × 79 = 474 7 × 79 = 553 8 × 79 = 632 9 × 79 = 711	113 79) 8927 79 102 −79 237 −237 0	Check: 113 × 79 1017 7910 8927

9.

a. $2 \times 10^4 = 20{,}000$	b. $712 \times 10^3 = 712{,}000$	c. $55 \times 10^6 = 55{,}000{,}000$
d. $6 \times 10^3 = 6{,}000$	e. $18 \times 10^7 = 180{,}000{,}000$	f. $69 \times 10^6 = 69{,}000{,}000$

10.

a. 15,278 × 3,892 (round to thousands)	b. 19,945,020 − 6,320,653 (round to millions)
My estimation: 15,000 × 4,000 = 60,000,000	My estimation: 20,000,000 − 6,000,000 = 14,000,000
Exact answer: 59,461,976	Exact answer: 13,624,367
Error of estimation: 538,024	Error of estimation: 375,633

Chapter 3 Review, p. 124

1. a. 211   b. 311

2. a. $x = (164 - 72) \div 2 = 46$

   b. $x = (1080 - 420) \div 5 = 132$

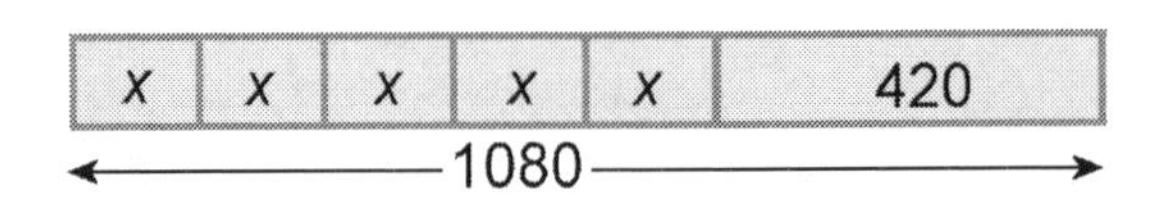

3. \$109 − \$25 − \$10 = \$74

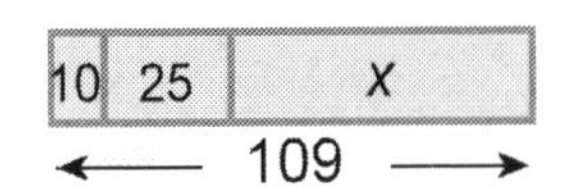

4. a. (137 qt − 45 qt) ÷ 2 = 46 qt. Eva canned 46 quarts.

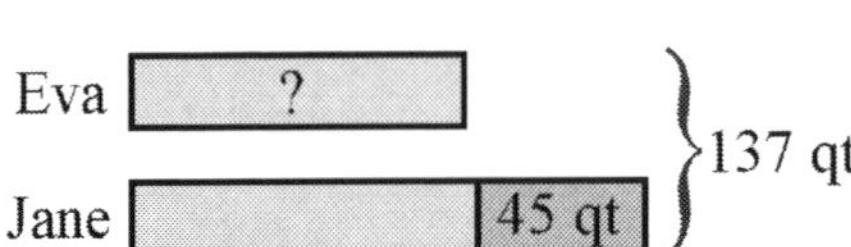

   b. Joe's raft was 3 ft longer than Jay's, which is 1/3 of Joe's raft.
   One block in the model is therefore 3 ft. Jay's raft was 6 feet long.

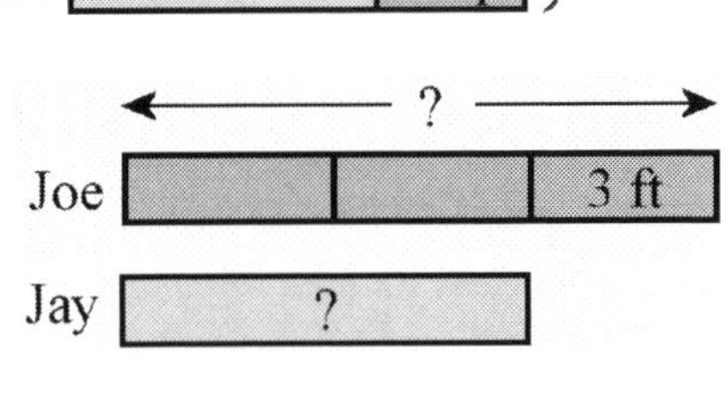

   c. (112 mi + 35 mi) × 2 = 294 mi

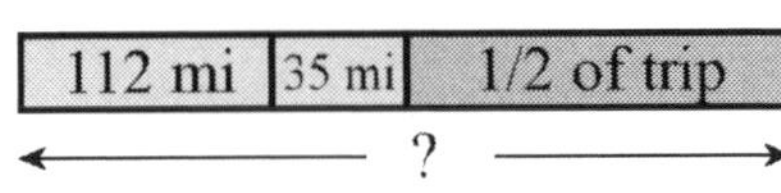

   d. One light bulb costs \$7.50 ÷ 5 = \$1.50. Eight would cost 8 × \$1.50 = \$12.00.

5. (\$175 − \$37) ÷ 2 = \$69. Austin was given \$69. \$69 + \$37 = \$106. Brandon was given \$106.

6. The discounted price is 3/4 of the normal price. Three-fourths of \$364 is \$364 ÷ 4 × 3 = \$273.

7. One block in the model is 69 ÷ 3 = 23. The chess club has 4 × 23 = 92 members.

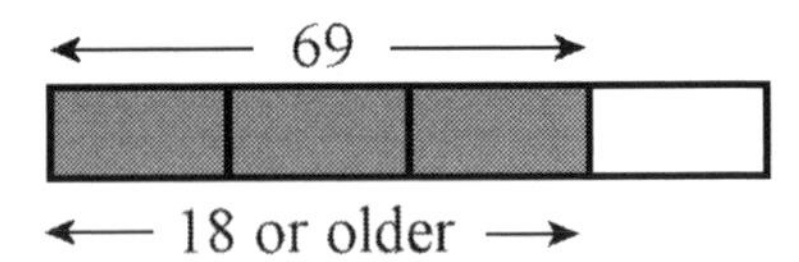

Chapter 4: Decimals

Review: Tenths and Hundredths, p. 130

1. a. $\frac{4}{10} = 0.4$ b. $\frac{9}{100} = 0.09$ c. $\frac{2}{10} = 0.2$ d. $\frac{78}{100} = 0.78$

 e. $1\frac{55}{100} = 1.55$ f. $\frac{27}{100} = 0.27$ g. $1\frac{9}{100} = 1.09$ h. $2\frac{78}{100} = 2.78$

2.	fraction/ mixed number	read as ...
a. 0.45	$\frac{45}{100}$	"forty-five hundredths" or " zero point four five"
b. 3.97	$3\frac{97}{100}$	"three and ninety-seven hundredths" or "three point nine seven"
c. 5.02	$5\frac{2}{100}$	"five and two hundredths" or "five point oh two"
d. 3.6	$3\frac{6}{10}$	"three and six tenths" or "three point six"
e. 12.60	$12\frac{60}{100}$	"twelve and sixty hundredths" or "twelve point six oh"

3. a. $0.5 + 0.6 = 1.1$ b. $0.9 + 0.5 = 1.4$
 c. $1.3 - 0.8 = 0.5$ d. $1.3 - 0.9 = 0.4$

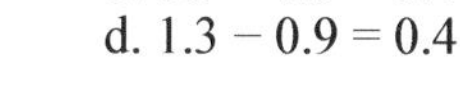

4.
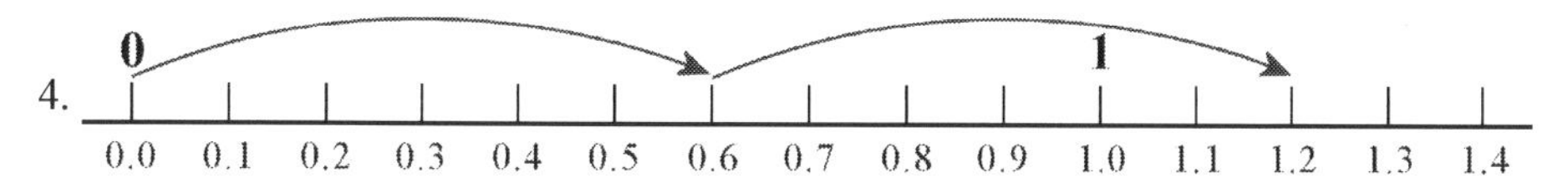

5.	decimal	fraction	read as ...
a.	0 . 8	$\frac{8}{10}$	eight tenths
	0 . 8 0	$\frac{80}{100}$	eighty hundredths
b.	1 . 2	$1\frac{2}{10}$	one and two tenths
	1 . 2 0	$1\frac{20}{100}$	one and twenty hundredths

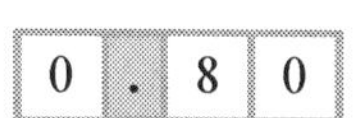

6.

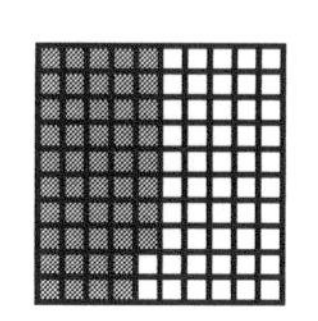

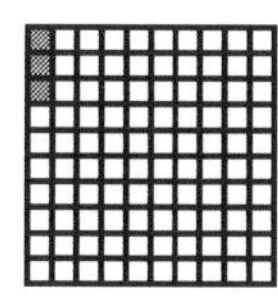

a. 0.48 b. 0.25 c. 0.03 d. 0.8

7. a. 0.4 b. 0.8 c. 0.6 d. 0.2 e. 0.78 f. 0.48 g. 0.22 h. 0.79

8. a. = b. < c. > d. = e. < f. < g. > h. > i. < j. >

More Decimals: Thousandths, p. 133

1.

a.

O		t	h	th
0	.	9	0	6

$= \frac{906}{1000}$

$= 9 \times \frac{1}{10} + 0 \times \frac{1}{100} + 6 \times \frac{1}{1000}$

b.

O		t	h	th
0	.	2	4	4

$= \frac{244}{1000}$

$= 2 \times \frac{1}{10} + 4 \times \frac{1}{100} + 4 \times \frac{1}{1000}$

c.

O		t	h	th
0	.	6	5	5

$= \frac{655}{1000}$

$= 6 \times \frac{1}{10} + 5 \times \frac{1}{100} + 5 \times \frac{1}{1000}$

d.

O		t	h	th
0	.	1	8	

$= \frac{18}{100}$

$= 1 \times \frac{1}{10} + 8 \times \frac{1}{100}$

e.

O		t	h	th
0	.	8	0	2

$= \frac{802}{1000}$

$= 8 \times \frac{1}{10} + 0 \times \frac{1}{100} + 2 \times \frac{1}{1000}$

f.

O		t	h	th
0	.	7	1	1

$= \frac{711}{1000}$

$= 7 \times \frac{1}{10} + 1 \times \frac{1}{100} + 1 \times \frac{1}{1000}$

2.

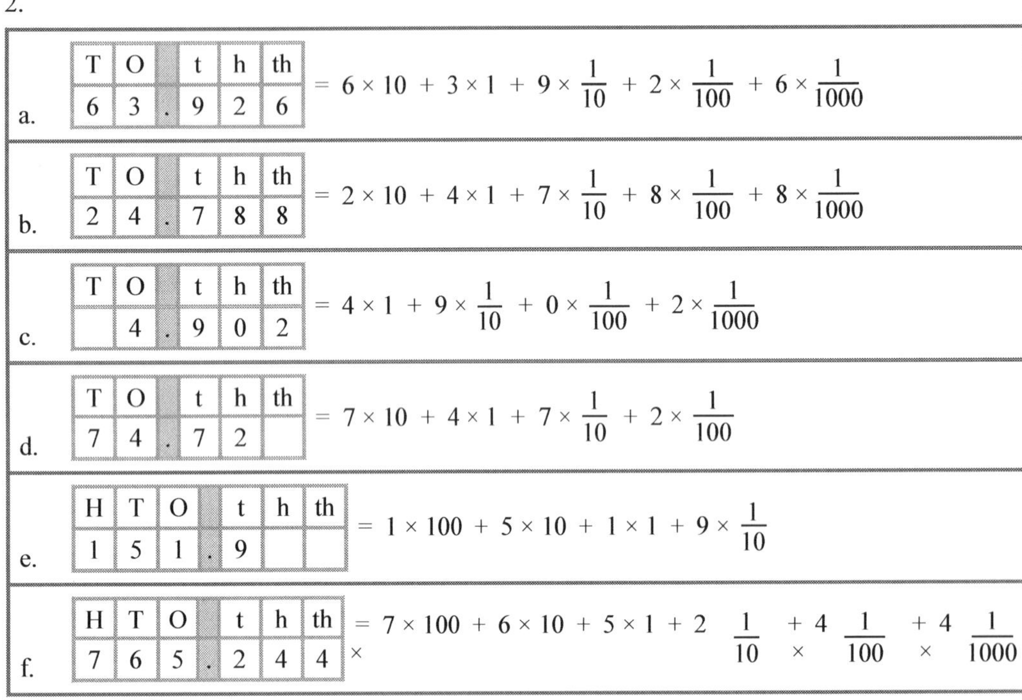

a.

T	O		t	h	th
6	3	.	9	2	6

$= 6 \times 10 + 3 \times 1 + 9 \times \frac{1}{10} + 2 \times \frac{1}{100} + 6 \times \frac{1}{1000}$

b.

T	O		t	h	th
2	4	.	7	8	8

$= 2 \times 10 + 4 \times 1 + 7 \times \frac{1}{10} + 8 \times \frac{1}{100} + 8 \times \frac{1}{1000}$

c.

T	O		t	h	th
	4	.	9	0	2

$= 4 \times 1 + 9 \times \frac{1}{10} + 0 \times \frac{1}{100} + 2 \times \frac{1}{1000}$

d.

T	O		t	h	th
7	4	.	7	2	

$= 7 \times 10 + 4 \times 1 + 7 \times \frac{1}{10} + 2 \times \frac{1}{100}$

e.

H	T	O		t	h	th
1	5	1	.	9		

$= 1 \times 100 + 5 \times 10 + 1 \times 1 + 9 \times \frac{1}{10}$

f.

H	T	O		t	h	th
7	6	5	.	2	4	4

$= 7 \times 100 + 6 \times 10 + 5 \times 1 + 2 \times \frac{1}{10} + 4 \times \frac{1}{100} + 4 \times \frac{1}{1000}$

3.

a.

O		t	h	th
0	.	0	0	7

$= \frac{7}{1000}$

b.

O		t	h	th
0	.	8	0	2

$= \frac{802}{1000}$

c.

O		t	h	th
3	.	3	7	1

$= 3\frac{371}{1000}$

d.

O		t	h	th
0	.	0	3	9

$= \frac{39}{1000}$

e.

O		t	h	th
1	.	4	1	

$= 1\frac{41}{100}$

f.

O		t	h	th
7	.	0	0	4

$= 7\frac{4}{1000}$

More Decimals: Thousandths, cont.

4. a. $0.95 = 9 \times (1/10) + 5 \times (1/100)$
   b. $1.405 = 1 \times 1 + 4 \times (1/10) + 0 \times (1/100) + 5 \times (1/1000)$
   c. $244.781 = 2 \times 100 + 4 \times 10 + 4 \times 1 + 7 \times (1/10) + 8 \times (1/100) + 1 \times (1/1000)$
   d. $65.05 = 6 \times 10 + 5 \times 1 + 0 \times (1/10) + 5 \times (1/100)$
   e. $20.214 = 2 \times 10 + 0 \times 1 + 2 \times (1/10) + 1 \times (1/100) + 4 \times (1/1000)$

5. When you divide one whole into ten equal parts, you get **tenths**.
   When you divide one tenth into ten equal parts, you get **hundredths**.
   When you divide one hundredth into ten equal parts, you get **thousandths**.

6. a. 0.003 (three thousandths) b. 0.007 (seven thousandths)
   c. 0.015 (fifteen thousandths) d. 0.018 (eighteen thousandths)
   e. 0.022 (twenty-two thousandths)

7.
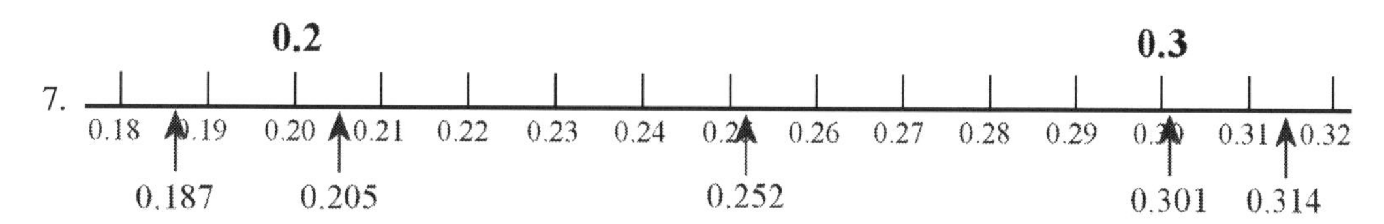

8. a. 0.003 b. 0.012 c. 0.319 d. 0.050
   e. 4.034 f. 2.004 g. 3.03 h. 1.80
   i. 17.003 j. 0.649 k. 9.01 l. 50.619

9. a. $0.048 = \frac{48}{1000}$ b. $3.902 = 3\frac{902}{1000}$ c. $3.005 = 3\frac{5}{1000}$ d. $6.7 = 6\frac{7}{10}$
   e. $10.06 = 10\frac{6}{100}$ f. $12.060 = 12\frac{60}{1000}$ g. $7.90 = 7\frac{90}{100}$ h. $0.429 = \frac{429}{1000}$
   i. $505.5 = 505\frac{5}{10}$ j. $4.789 = 4\frac{789}{1000}$ k. $0.091 = \frac{91}{1000}$ l. $5.42 = 5\frac{42}{100}$

10. a. 0.273 b. 1.89 c. 7.203 d. 90.528 e. 17.108 f. 205.008

11. a. 10 thousandths makes a hundredth. $10 \times 0.001 = 0.01$
    b. 10 hundredths makes a tenth. $10 \times 0.01 = 0.1$
    c. 100 thousandths makes a tenth. $100 \times 0.001 = 0.1$
    d. 100 hundredths makes one whole. $100 \times 0.01 = 1$

12.

a.

O	.	t	h	th
0	.	2	8	5

one tenth more than 0.285 **0.385**
one hundredth more than 0.285 **0.295**
one thousandth more than 0.285 **0.286**

b.

O	.	t	h	th
0	.	0	1	6

one tenth more than 0.016 **0.116**
one hundredth more than 0.016 **0.026**
one thousandth more than 0.016 **0.017**

c.

O	.	t	h	th
1	.	0	7	

one tenth more than 1.07 **1.17**
one hundredth more than 1.07 **1.08**
one thousandth more than 1.07 **1.071**

d.

O	.	t	h	th
0	.	9		

one tenth more than 0.9 **1.0**
one hundredth more than 0.9 **0.91**
one thousandth more than 0.9 **0.901**

e.

O	.	t	h	th
2	.	3	1	6

five tenths more than 2.316 **2.816**
six hundredths more than 2.316 **2.376**
two thousandths more than 2.316 **2.318**

f.

T	O	.	t	h	th
	1	.	0	8	

ten more than 1.08 **11.08**
one hundredth more than 1.08 **1.09**
nine thousandths more than 1.08 **1.089**

More Decimals: Thousandths, cont.

13.

a. 2.	0	4	9		b. 3.
0					0
9			c. 2.	0	7
	d. 0.	0	5		6
	0				
	0		e. 0.	7	1
f. 0.	3	9	2		

Comparing Decimals, p. 138

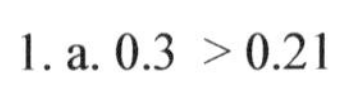

1. a. 0.3 > 0.21

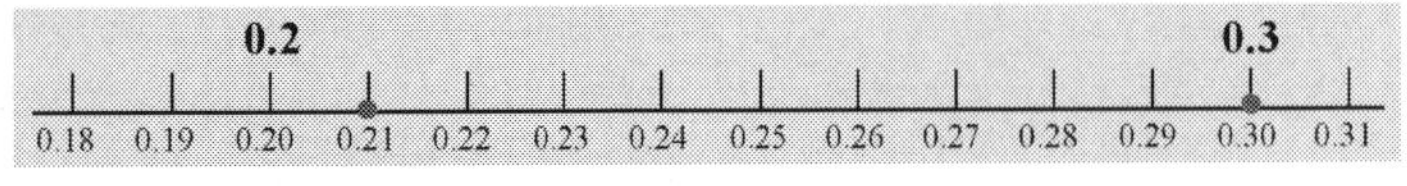

b. 5.02 < 5.2

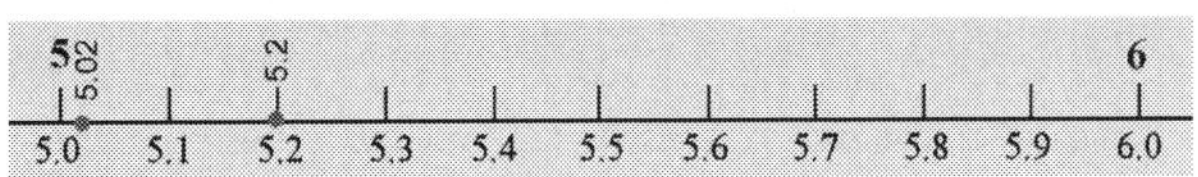

2. 4.8 < 4.92 < 5.01 < 5.03 < 5.1 < 5.15 < 5.19 < 5.24 < 5.3

3. a. 0.6 > 0.006 b. 0.03 < 0.3 c. 0.8 > 0.008 d. 0.80 = 0.800

4. a. 2.100 > 2.009 b. 4.004 < 4.010 c. 5.007 < 5.400 d. 3.004 < 3.400
e. 3.005 < 3.500 f. 6.001 < 6.010 g. 3.002 < 3.020 h. 7.060 < 7.606

5. a. 7.892 b. 15.404 c. 2.377 d. 3.99 e. 0.366 f. 0.4

6. a. 5.006, 5.06, 5.066, 5.6, 5.606, 5.66 b. 7.7, 7.708, 7.77, 7.78, 7.8, 7.807

7.

0.003	0.052	0.04	0.03	0.002	0.504	0.405	0.501	0.506
0.005	0.011	0.019	0.02	0.001	0.15	0.3	0.45	0.459
0.004	0.013	0.01	0.04	0.01	0.1	0.078	0.099	0.46
0.007	0.008	0.046	0.045	0.06	0.093	0.12	0.05	0.463
0.14	0.225	0.038	0.46	0.068	0.111	0.108	0.469	0.477
0.109	0.87	0.843	0.8	0.245	0.109	0.6	0.56	0.5
0.191	0.9	0.42	0.684	0.63	0.619	0.606	0.506	0.55
0.96	0.931	0.943	0.8	0.74	0.62	0.616	0.602	0.556
0.981	0.903	0.934	0.548	0.359	0.298	0.56	0.483	0.538

Rounding, p. 140

1. a. 2.894 ≈ 2.9 b. 206.8 ≈ 210 c. 36.92 ≈ 37 d. 408.517 ≈ 408.5

   e. 2.263 ≈ 2.3 f. 14.82 ≈ 10 g. 35.77 ≈ 36 h. 98.382 ≈ 98.38

2. a. 3.6 b. 55.9 c. 2.9 d. 210.2

3.

This rounded to the nearest →	unit (one)	tenth	hundredth
5.125	5	5.1	5.13
5.170	5	5.2	5.17
5.467	5	5.5	5.47
5.934	6	5.9	5.93

This rounded to the nearest →	unit (one)	tenth	hundredth
7.073	7	7.1	7.07
12.390	12	12.4	12.39
76.103	76	76.1	76.10
0.957	1	1.0	0.96

4. a. 3.964 ≈ 4.0 b. 4.928 ≈ 4.9 c. 15.973 ≈ 16.0 d. 209.98 ≈ 210.0

5. a. 5.296 ≈ 5.30 b. 2.998 ≈ 3.00 c. 76.297 ≈ 76.30 d. 39.998 ≈ 40.00

6.

Round to the nearest ...	0.709	14.097	12.995	44.385	2.905	79.992
... whole number	1	14	13	44	3	80
... tenth	0.7	14.1	13.0	44.4	2.9	80.0
... hundredth	0.71	14.10	13.00	44.39	2.91	79.99

7.

a. 2	a. 5.	6	d. 1	
	5		d. 2.	3
	b. 4	9.	9	
				e. 0.
c. 1	5	b. 8.	4	0
3		4		6

Add and Subtract Decimals, p. 142

1. a.

O	.	t	h	th
0	.	2	8	5

1 tenth more: 0.385
1 hundredth more: 0.295
1 thousandth more: 0.286

b.

O	.	t	h	th
0	.	0	1	6

2 tenths more: 0.216
2 hundredths more: 0.036
2 thousandths more: 0.018

c.

O	.	t	h	th
1	.	0	7	

5 tenths more: 1.57
2 hundredths more: 1.09
6 thousandths more: 1.076

Add and Subtract Decimals, cont.

2.

a. 0.009 + 0.006 = 0.015 0.009 + 0.060 = 0.069 0.009 + 0.600 = 0.609	b. 0.8 + 0.6 = 1.4 0.80 + 0.06 = 0.86 0.800 + 0.006 = 0.806

3.

a. 0.700 + 0.005 = 0.705 0.07 + 0.50 = 0.57 0.007 + 0.050 = 0.057	b. 0.7 − 0.5 = 0.2 0.70 − 0.07 = 0.63 0.050 − 0.007 = 0.043
c. 0.2 + 0.9 = 1.1 0.02 + 0.90 = 0.92 0.020 + 0.009 = 0.029	d. 0.90 − 0.05 = 0.85 0.77 − 0.03 = 0.74 0.770 − 0.003 = 0.767

4. a. Laura should think in *thousandth parts*:
8 thousandths + 3 thousandths = 11 thousandths.
0.008 + 0.003 = 0.011

b. Jessie should think in *tenth parts*:
7 tenths + 7 tenths = 14 tenths = 1 whole 4 tenths
0.7 + 0.7 = 1.4

5.

a. 0.6 + 0.4 = 1 b. 0.60 + 0.40 = 1 c. 0.61 + 0.39 = 1	d. 0.99 + 0.01 = 1 e. 0.87 + 0.13 = 1 f. 0.22 + 0.78 = 1	g. 0.999 + 0.001 = 1 h. 0.002 + 0.998 = 1 i. 0.304 + 0.696 = 1

6.

a. 1 − 0.01 = 0.99 b. 1 − 0.04 = 0.96 c. 1 − 0.51 = 0.49	d. 1 − 0.001 = 0.999 e. 1 − 0.008 = 0.992 f. 1 − 0.021 = 0.979	g. 1 − 0.506 = 0.494 h. 1 − 0.56 = 0.44 i. 1 − 0.411 = 0.589

7.

a. $0.50 + x = 0.677$ $x = 0.177$	b. $x + 1.52 = 2$ $x = 0.48$	c. $1 - x = 0.378$ $x = 0.622$	d. $x - 0.5 = 1.27$ $x = 1.77$

8.

```
a.                 b.                    c.     12 9
                                               4 2 10 10
   1 4 5 . 5 0        1 9 0 . 0 0 0          2 5 3 . 0 0
 −   2 4 . 9 3        3 4 2 . 2 5 0        −   2 3 . 2 8
 -------------    +     4 5 . 8 0 8        -------------
   1 2 0 . 5 7    -----------------          2 2 9 . 7 2
                      5 7 8 . 0 5 8
```

9. a. 0.56 + 0.39 = 0.95 ≈ 1.0   b. 1.09 − 0.549 = 0.541 ≈ 0.5

10.

	− 0.06		− 0.08		− 0.1		− 0.04		− 0.2		− 0.07		− 0.02		− 0.03	
8.28		8.22		8.14		8.04		8.00		7.8		7.73		7.71		7.68

11. The puppy had gained 0.134 kg.

12. a. The total weight was 6.325.   b. It was 0.675 kg less than 7 kg.

13. a. 4.025 miles   b. 3.428 miles

Multiplying Decimals by Whole Numbers, p. 146

1.

a. $2 \times 0.8 = 0.8 + 0.8 = 1.6$	b. $3 \times 1.5 = 1.5 + 1.5 + 1.5 = 4.5$
c. $4 \times 0.03 = 0.03 + 0.03 + 0.03 + 0.03 = 0.12$	d. $2 \times 0.007 = 0.007 + 0.007 = 0.014$

2.

a. $10 \times 0.4 = 4.0$ b. $100 \times 0.4 = 40.0$ c. $1000 \times 0.4 = 400.0$	g. $7 \times 0.05 = 0.35$ h. $8 \times 0.05 = 0.40$ i. $10 \times 0.05 = 0.50$	m. $4 \times 0.004 = 0.016$ n. $5 \times 0.008 = 0.040$ o. $3 \times 0.012 = 0.036$
d. $8 \times 0.4 = 3.2$ e. $80 \times 0.4 = 32.0$ f. $800 \times 0.4 = 320.0$	j. $10 \times 0.09 = 0.90$ k. $100 \times 0.09 = 9.00$ l. $1000 \times 0.09 = 90.00$	p. $10 \times 0.003 = 0.030$ q. $100 \times 0.003 = 0.300$ r. $1000 \times 0.003 = 3.000$

3.

a.	b.	c.
$7 \times 0.01 = 0.07$	$10 \times 0.1 = 1.0$	$1 \times 0.005 = 0.005$
$7 \times 0.02 = 0.14$	$10 \times 0.2 = 2.0$	$2 \times 0.005 = 0.010$
$7 \times 0.03 = 0.21$	$10 \times 0.3 = 3.0$	$3 \times 0.005 = 0.015$
$7 \times 0.04 = 0.28$	$10 \times 0.4 = 4.0$	$4 \times 0.005 = 0.020$
$7 \times 0.05 = 0.35$	$10 \times 0.5 = 5.0$	$5 \times 0.005 = 0.025$
$7 \times 0.06 = 0.42$	$10 \times 0.6 = 6.0$	$6 \times 0.005 = 0.030$
$7 \times 0.07 = 0.49$	$10 \times 0.7 = 7.0$	$7 \times 0.005 = 0.035$
$7 \times 0.08 = 0.56$	$10 \times 0.8 = 8.0$	$8 \times 0.005 = 0.040$
$7 \times 0.09 = 0.63$	$10 \times 0.9 = 9.0$	$9 \times 0.005 = 0.045$

4.

a. $3 \times 0.7 = 2.1$ b. $3 \times 0.07 = 0.21$ c. $3 \times 0.007 = 0.021$	d. $12 \times 0.5 = 6.0$ e. $12 \times 0.05 = 0.60$ f. $12 \times 0.005 = 0.060$	g. $10 \times 0.4 = 4.0$ h. $100 \times 0.04 = 4.00$ i. $1000 \times 0.004 = 4.000$

5.

a.	b.	c.
$5 \times 100 = 500$	$6 \times 400 = 2400$	$9 \times 800 = 7200$
$5 \times 10 = 50$	$6 \times 40 = 240$	$9 \times 80 = 720$
$5 \times 1 = 5$	$6 \times 4 = 24$	$9 \times 8 = 72$
$5 \times 0.1 = 0.5$	$6 \times 0.4 = 2.4$	$9 \times 0.8 = 7.2$
$5 \times 0.01 = 0.05$	$6 \times 0.04 = 0.24$	$9 \times 0.08 = 0.72$
$5 \times 0.001 = 0.005$	$6 \times 0.004 = 0.024$	$9 \times 0.008 = 0.072$

6. a. Lucy jogs 5 × 1.2 mi = 6 miles. Sharon jogs 4 × 1.5 = 6 miles.
   They both job the same distance during the week.

   b. One-half is 1.5, which is less than 1.6, because six tenths is more than five tenths.

   c. Five kilograms is 5 × 2.2 lb = 11 lb.

   d. The total height of ten books is 10 × 0.8 cm = 8 cm. Yes, you can stack 10 books in a 15-centimeter high box.
   The total height of 12 books is 12 × 0.8 cm = 9.6 cm, so, yes, you can stack 12 books in the box.
   The maximum number of books is 18. A stack of 20 books would be 16 cm high. A stack of 19 books would be 15.2 cm tall. So, 18 books fit.

Multiplying Decimals by Whole Numbers, cont.

7.

a. 0.	2	4		b. 0.	9
0				2	
7			c. 0.	7	0
2			4		
	d. 0.	9	9	0	
	8				e. 3.
	0		e. 5.	0	0

8. a. $4 \times 0.5 = 2$ b. $12 \times 0.05 = 0.6$ c. $10 \times 0.03 = 0.3$
d. $10 \times 0.009 = 0.09$ e. $100 \times 0.002 = 0.2$ f. $5 \times 0.06 = 0.3$

9.

0.04
40×0.001

4
10×0.4 8×0.5 20×0.2

0.4
40×0.01 10×0.04 8×0.05 20×0.02

10. *How did Mrs. Decimal feel when she lost her decimal point?*

11.1	16.56	10.11		21.6	11.01		30.4	34	31	30.012	3.04
O	U	T		O	F		P	L	A	C	E

Multiplying Decimals in Columns, p. 150

1.

a. $8 \times 13.1 = 104.8$ $8 \times 1.31 = 10.48$	b. $15 \times 5.62 = 84.30$ $15 \times 56.2 = 843.0$	c. $22 \times 8.06 = 177.32$ $2.2 \times 806 = 1773.2$

2. a. 16.2 b. 56.203 c. 37.44 d. 136.8 e. 5.465 f. $6.16

3. Four books. Use estimation first: 4×2.5 kg = 10 kg. Then check: 4×2.35 kg = 9.4 kg. We could not add a fifth book without the weight going over 10 kg.

4. 3×0.478 L = 1.434 L, so, you have 0.434 liters more than one liter.

5. $2 \times 198 \times 0.78 = 308.88$ km

6. Twelve meters would cost $6 \times \$2.72 = \$16.32$..

7. The 2-gallon container holds 8 quarts. Guess and check. 5×1.6 qt = 8 qt. You can fill the gas tank five times.

Puzzle Corner: a. The answer must be near $4 \times 7 = 28$, so it is 27.122
b. The answer must be near $5 \times 3 = 15$, so it is 13.905
c. The answer must be near $6 \times 3 = 18$, so it must be 15.4199

Multiplying Decimals by Decimals, p. 152

1.

a. one-tenth of 50 $0.1 \times 50 = 5$ b. three-tenths of 50 $0.3 \times 50 = 15$	c. one-tenth of 700 $0.1 \times 700 = 70$ d. four-tenths of 700 $0.4 \times 700 = 280$	e. one-hundredth of 4,000 $0.01 \times 4000 = 40$ f. six-hundredths of 4,000 $0.06 \times 4000 = 240$

2.

a. $0.1 \times 30 = 3$ $0.4 \times 30 = 12$	b. $0.1 \times 400 = 40$ $0.6 \times 400 = 240$	c. $0.01 \times 600 = 6$ $0.07 \times 600 = 42$
d. $0.1 \times 520 = 52$ $0.3 \times 520 = 156$	e. $0.001 \times 5{,}000 = 5$ $0.002 \times 5{,}000 = 10$	f. $0.01 \times 800 = 8$ $0.11 \times 800 = 88$

3. a. The result will be less.
   b. The result will be less.
   c. The result will be more.

4.

a. 0.1×40 px = 4 px	b. 0.3×40 px = 12 px	c. 0.6×40 px = 24 px
d. 0.2×40 px = 8 px	e. 0.5×40 px = 20 px	f. 0.9×40 px = 36 px

5.

a. 0.5×50 px = 25 px	b. 0.3×50 px = 15 px
c. 1.5×50 px = 75 px	d. 1.3×50 px = 65 px

6. a. longer b. shorter c. shorter d. longer

7.

a. To calculate 0.8×0.8, I first multiply $8 \times 8 = 64$. The answer to 0.8×0.8 has to be *slightly smaller* than 0.8, because scaling anything by 0.8 is close to the original, but somewhat smaller. So, 0.8×0.8 cannot be 64, and it cannot be 6.4, but it is 0.64 !
b. 0.1×5.6 has to be 1/10 of the size of 5.6. So, it cannot be 56. Could it be 5.6? No, because $1 \times 5.6 = 5.6$. So, 0.1×5.6 has to equal 0.56 .
c. 0.4×0.06 has to be smaller than 0.06. It can be neither 24, nor 2.4. Is it 0.24 or 0.024? 0.024

8.

a. $0.5 \times 0.3 = 0.15$ b. $0.9 \times 0.6 = 0.54$	c. $0.4 \times 0.08 = 0.032$ d. $0.7 \times 0.02 = 0.014$	e. $0.1 \times 0.3 = 0.03$ f. $0.1 \times 2.7 = 0.27$
g. $0.2 \times 0.1 = 0.02$ h. $0.8 \times 0.1 = 0.08$	i. $0.9 \times 0.01 = 0.009$ j. $0.9 \times 0.1 = 0.09$	k. $0.7 \times 0.3 = 0.21$ l. $7 \times 0.03 = 0.21$

9.

a. $0.4 \times 0.8 = 0.32$ b. $0.7 \times 1.1 = 0.77$ c. $0.02 \times 0.9 = 0.018$	d. $0.02 \times 0.5 = 0.01$ e. $0.002 \times 9 = 0.018$ f. $1.1 \times 0.3 = 0.33$	g. $2.1 \times 0.2 \times 0.5 = 0.21$ h. $0.4 \times 4 \times 0.2 = 0.32$ i. $6 \times 0.06 \times 0.2 = 0.072$

Multiplying Decimals by Decimals, cont.

10.

a. $0.4 \times 0.5 = 0.2$	d. $3 \times 0.2 \times 0.5 = 0.3$	g. $0.6 \times 0.2 \times 0.5 = 0.06$
b. $20 \times 0.06 = 1.2$	e. $300 \times 0.009 = 2.7$	h. $600 \times 0.004 = 2.4$
c. $40 \times 0.05 = 2$	f. $40 \times 0.05 = 2$	i. $0.4 \times 0.5 \times 60 = 12$

11. a. The ribbon costs $0.4 \times \$1.10 = \$0.44$.
    b. The nuts cost $0.3 \times \$8.00 = \$2.40$.
    c. The phone call costs $1.2 \times \$7.00 = \$8.40$.
    d. The lace costs $1.5 \times \$2.20 = \$3.30$.

More Decimal Multiplication, p. 156

1.

a. $4 \times 0.04 = 0.16$	e. $0.7 \times 0.9 = 0.63$	i. $3 \times 0.21 = 0.63$
b. $4 \times 0.004 = 0.016$	f. $9 \times 0.007 = 0.063$	j. $0.3 \times 0.21 = 0.063$
c. $0.4 \times 0.4 = 0.16$	g. $0.07 \times 0.9 = 0.063$	k. $2.1 \times 0.3 = 0.63$
d. $0.4 \times 0.04 = 0.016$	h. $7 \times 0.09 = 0.63$	l. $21 \times 0.003 = 0.063$

2.

a. $0.5 \times 0.7 = 0.35$	d. $4 \times 0.6 = 2.4$	g. $0.9 \times 8 = 7.2$
b. $0.5 \times 0.07 = 0.035$	e. $4 \times 0.006 = 0.024$	h. $0.9 \times 0.09 = 0.081$
c. $0.05 \times 0.7 = 0.035$	f. $4 \times 0.06 = 0.24$	i. $0.9 \times 0.6 = 0.54$

3. Answers will vary. Check the students' work. Examples:

a. $0.6 \times 0.4 = 0.24$	c. $0.06 \times 0.7 = 0.042$	e. $9 \times 0.06 = 0.54$
b. $2 \times 0.12 = 0.24$	d. $7 \times 0.006 = 0.042$	f. $0.6 \times 0.9 = 0.54$

4. Answers will vary. Please check the students' work. Examples:
   $0.04 \times 7 = 0.28$ $4 \times 0.07 = 0.28$ $0.4 \times 0.7 = 0.28$

5.

Across:

a. $7.1 \times 0.01 = 0.071$
b. $0.9 \times 1.1 = 0.99$
d. $0.5 \times 1.1 = 0.55$
e. $1.5 \times 0.04 = 0.06$
f. $10 \times 0.333 = 3.33$

Down:

a. $9 \times 0.101 = 0.909$
b. $0.05 \times 0.5 = 0.025$
c. $0.8 \times 0.08 = 0.064$
d. $0.2 \times 3.4 = 0.68$
e. $9 \times 0.07 = 0.63$

			a. 0.	0	7	1
	b. 0.	9	9			
	0		0			c. 0.
	2		9			0
d. 0.	5	5		e. 0.	0	6
6				6		4
8		f. 3.	3	3		

6. a. 0.3×6 kg = 1.8 kg   b. 0.8×0.15 L = 0.12 L   c. 0.11×2 lb = 0.22 lb

7. It would cost \$0.63. The discount is 3/10 of \$0.90, or $0.3 \times \$0.90 = \$0.27$. The new price is $\$0.90 - \$0.27 = \$0.63$.
   You can also calculate the new price as being 7/10 of \$0.90, or $0.7 \times \$0.90 = \$0.63$.

8. It would cost \$9. The discount is 4/10 of \$15, or $0.4 \times \$15 = \$6$. The new price is $\$15 - \$6 = \$9$.
   You can also calculate the new price as being 6/10 of \$15, or $0.6 \times \$15 = \$9$.

9. He pays $0.11 \times \$2{,}000 = \$220$.

10. 0.6×2.1 mi = 1.26 m. The item will be on the track 1.26 miles from the start.

More Decimal Multiplication, cont.

11.

a. $0.5 \times 4 \times 0.8 = 1.6$ b. $50 \times 0.6 \times 1.1 = 33$ c. $0.18 \times 0.2 \times 10 = 0.36$	d. $300 \times 0.002 = 0.6$ e. $4{,}000 \times 0.03 = 120$ f. $0.009 \times 20 \times 200 = 36$

12.

a. $(0.2 + 0.9) \times 0.6 = 0.66$	c. $56 \div 8 \times 0.7 = 4.9$
b. $0.2 + 0.9 \times 0.6 = 0.74$	d. $3 - 0.9 \times (1 - 0.7) = 2.73$

13. The new width is 1.4×5 in. = 7 inches.
The new height is 1.4×2 in. = 2.8 inches.

14. Bleach: 0.06×3 L = 0.18 L.
Water: 3 L − 0.18 L = 2.82 L.

15.

a. $1.2 \times 0.435 > 0.435$	b. $2.03 \times 0.39 > 0.39$	c. $0.49 \times 0.27 < 0.27$
d. $0.8 \times 0.435 < 0.435$	e. $1.05 \times 0.39 > 0.39$	f. $0.49 \times 0.27 < 0.49$

Puzzle Corner:
$9 \times 12.3 = 110.7$; $9 \times 12.34 = 111.06$; $9 \times 12.345 = 111.105$
Guess: 9×12.3456 would be 111.1104. Checking with a calculator: yes, it is.
$9 \times 12.34567 = 111.11103$, so the pattern continued here.
$9 \times 12.345678 = 111.111102$, so the pattern still continued. (What about 9×12.3456789 ?)

Long Multiplication, p. 159

1. a. $5.69 \times 14.7 = 83.643$ b. $0.381 \times 1.54 = 0.58674$ c. $0.041 \times 3.032 = 0.124312$

2.

a. Estimate: $0.3 \times 1 = 0.3$ Exact: $0.3 \times 1.19 = 0.357$	b. Estimate: $6 \times 3 = 18$ Exact: $5.6 \times 2.8 = 15.68$	c. Estimate: $3 \times 4 = 12$ Exact: $3.34 \times 4.2 = 14.028$

3. Riddle: *What did number 22 say to 21.21?* "*YOU'RE...*

2.736	0.424	0.42	2.415	3.28	2.25		0.14	1.6	2.16	3.553	3.5	0.12	0.4	1.2
B	E	S	I	D	E		Y	O	U	R	S	E	L	F

Dividing Decimals—Mental Math, p. 160

1.

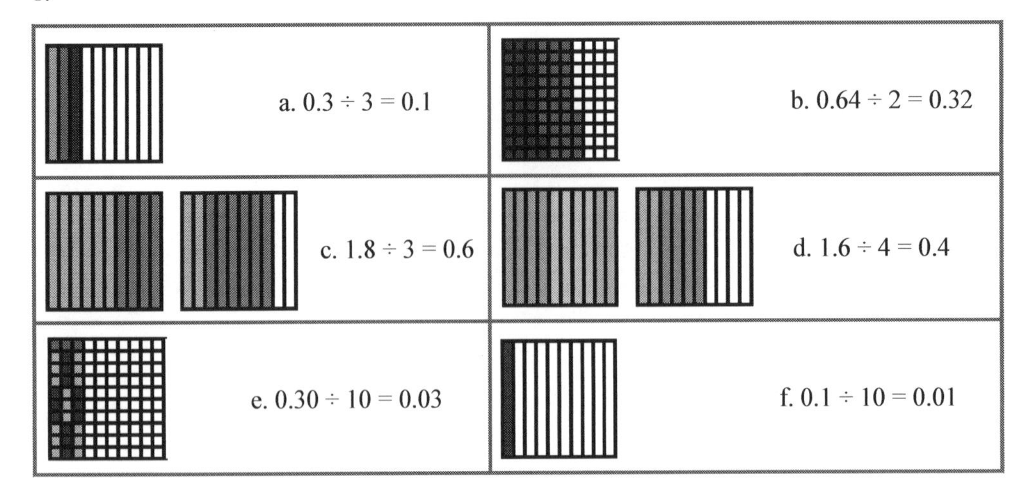

2. a. 9 tenths divided by 3 equals ...	0.9 ÷ 3 = 0.3
b. 72 thousandths divided by 9 equals ...	0.072 ÷ 9 = 0.008
c. 54 hundredths divided by 6 equals ...	0.54 ÷ 6 = 0.09
d. 240 thousandths divided by 60 equals ...	0.240 ÷ 60 = 0.004
e. 122 hundredths divided by 2 equals ...	1.22 ÷ 2 = 0.61

3.

a. 0.024 ÷ 6 = 0.004	d. 0.49 ÷ 7 = 0.07	g. 5.40 ÷ 9 = 0.6
b. 0.24 ÷ 6 = 0.04	e. 1.2 ÷ 3 = 0.4	h. 0.20 ÷ 4 = 0.05
c. 2.4 ÷ 6 = 0.4	f. 0.056 ÷ 7 = 0.008	i. 0.050 ÷ 10 = 0.005

4.

a. 0.30 ÷ 5 = 0.06	d. 0.060 ÷ 12 = 0.005	g. 0.300 ÷ 50 = 0.006
b. 0.30 ÷ 10 = 0.03	e. 0.200 ÷ 40 = 0.005	h. 0.700 ÷ 100 = 0.007
c. 3.0 ÷ 5 = 0.6	f. 2.0 ÷ 5 = 0.4	i. 0.020 ÷ 10 = 0.002

5. Each friend gets \$2.00 ÷ 5 = \$0.40.

6. 5 Five heartbeats take 5 × 0.8 seconds = 4 seconds, and ten heartbeats take 10 × 0.8 seconds = 8 seconds.

7.

a. 8 × 0.04 = 0.32	b. 0.4 × 5 = 2	c. 0.005 × 5 = 0.025
0.04 × 8 = 0.32	5 × 0.4 = 2	5 × 0.005 = 0.025
0.32 ÷ 8 = 0.04	2 ÷ 0.4 = 5	0.025 ÷ 0.005 = 5
0.32 ÷ 0.04 = 8	2 ÷ 5 = 0.4	0.025 ÷ 5 = 0.005

8.

a. 4.5 ÷ 0.5 = 9	d. 0.12 ÷ 0.06 = 2	g. 2.1 ÷ 0.7 = 3
b. 0.45 ÷ 0.05 = 9	e. 0.006 ÷ 0.002 = 3	h. 1.5 ÷ 0.3 = 5
c. 0.450 ÷ 0.005 = 90	f. 0.63 ÷ 0.07 = 9	i. 0.09 ÷ 0.01 = 9

9. a. 1.8 m ÷ 0.3 m = 6 pieces b. 4.2 m ÷ 0.7 m = 6 pieces c. 0.25 m ÷ 0.05 m = 5 pieces

Dividing Decimals—Mental Math, cont.

10.

a. 0.20 ÷ 0.05 = 4 d. 0.30 ÷ 0.05 = 6	b. 1.0 ÷ 0.2 = 5 e. 5.0 ÷ 0.2 = 25	c. 0.40 ÷ 0.02 = 20 f. 0.050 ÷ 0.001 = 50
g. 0.60 ÷ 0.05 = 12 j. 1.0 ÷ 0.02 = 50	h. 0.90 ÷ 0.01 = 90 k. 1.0 ÷ 0.01 = 100	i. 0.10 ÷ 0.01 = 10 l. 0.030 ÷ 0.002 = 15

11. a. 6 mi ÷ 1.2 mi/day = 5 days b. 60 mi ÷ 1.2 mi/day = 50 days

12. a. Jack has 29 nickels. b. $1.45 ÷ $0.05 = 29 nickels.

13. 0.20 m ÷ 0.04 m = 5 sticks.

14. 8 × 1.5 cm + 3 cm = 15 cm and (15 cm − 3 cm) ÷ 1.5 cm = 8

15. 5 m − 4 × 0.6 m = 2.6 m

16. He can get 2 servings, because 2 × 0.3 kg = 0.6 kg, but 3 × 0.3 kg = 0.9 kg, which is more than 0.85 kg. In grams, he can get two 300-gram servings out of 850 grams.

17.

		a. 1		d. 0.	5
	a. 2	5		0	
			c. 1	1	
b. 0.	0	0	2		e. 0.
0				e. 1	0
3					4

18.

a. 0.025 ÷ 0.005 = 5 0.25 ÷ 0.05 = 5 2.5 ÷ 0.5 = 5 25 ÷ 5 = 5 250 ÷ 50 = 5	b. 1000 ÷ 20 = 50 100 ÷ 2 = 50 10 ÷ 0.2 = 50 1 ÷ 0.02 = 50 0.1 ÷ 0.002 = 50	c. 4200 ÷ 40 = 105 420 ÷ 4 = 105 42 ÷ 0.4 = 105 4.2 ÷ 0.04 = 105 0.42 ÷ 0.004 = 105

Puzzle Corner: 987 ÷ 21 = 47

Long Division with Decimals, p. 165

1. a. 1.06 Check: 5 × 1.06 = 5.30   b. 0.24 Check: 3 × 0.24 = 0.72
   c. 0.89 Check: 7 × 0.89 = 6.23   d. 0.398 Check: 6 × 0.398 = 2.388

2. a. 1.26 Check: 19 × 1.26 = 23.94   b. 2.506 Check: 23 × 2.506 = 57.638

3. a. "First subtract $2.30 from $10.70. Then divide that result by 24. One roll costs $0.35."
   b. ($10.70 − $2.30) ÷ 24 = $0.35

4. One muffin costs ($7.11 – $1.23) ÷ 7 = $0.84.

5.

a. 31 ÷ 4 = 7 R3 31.00 ÷ 4 = 7.75 Check: 4 × 7.75 = 31.00	b. 56 ÷ 5 = 11 R1 56.0 ÷ 5 = 11.2 Check: 5 × 11.2 = 56.0	c. 15 ÷ 8 = 1 R7 15.000 ÷ 8 = 1.875 Check: 8 × 1.875 = 15.000	d. 45 ÷ 20 = 2 R5 45.00 ÷ 20 = 2.25 Check: 20 × 2.25 = 45.00

6. They are 0.29 miles long. Long division to three decimals gives 2.000 ÷ 7 = 0.285.

7. 0.25 kg or 250 g. Long division: 1.50 ÷ 6 = 0.25.

8. The average price is $2.74. Calculation: ($2.55 + $2.69 + $2.95 + $2.75) ÷ 4 = $10.94 ÷ 4 = $2.735

9. a. 0.130 ÷ 4 × 4 = 0.098 kg   b. 23 ÷ 5 × 3 = 13.8 seconds

More Long Division with Decimals, p. 169

1.

a. $\frac{5}{8} = 0.625$	b. $\frac{6}{7} = 0.857$	c. $\frac{1}{6} = 0.167$	d. $\frac{7}{20} = 0.350$
0.625 8) 5.000 -4 8 2 0 -1 6 4 0 - 4 0 0	0.8571 7) 6.0000 -5 6 4 0 -3 5 5 0 - 4 9 1 0 - 1 0 1	0.1666 6) 1.0000 -0 6 4 0 -3 6 4 0 - 3 6 4 0 - 3 6 4	0.350 20) 7.000 -6 0 1 0 0 -1 0 0 0

2. a. 41.67. The long division is 250.000 ÷ 6 = 41.666.   b. 3.409. The long division is 37.5000 ÷ 11 = 3.4090

3. First subtract the cost of seven capacitor packs from $8.70. Then divide that result by three.
   b. ($8.70 − 7 × 0.60) ÷ 3
   c. $1.50

4. Each person paid ($35.40 + $128.95) ÷ 3 = $54.78.

5. Answers will vary. Check the students' work. For example:
   a. Four people shared the cost of a $50 taxi ride so that one person paid $26 and the other three shared the remaining cost equally. How much did each of the three persons pay?

   b. Harry and Larry shared equally the cost of buying 25 cheap CDs for $1.40 each. How much did each person pay?

Multiply and Divide by Powers of Ten, p. 172

1. a. $10 \times 0.04 = 0.4$
   b. $100 \times 0.04 = 4$
   c. $1000 \times 0.04 = 40$
   d. $10 \times 0.56 = 5.6$
   e. $100 \times 0.56 = 56$
   f. $000 \times 0.56 = 560$
   g. $10 \times 0.048 = 0.48$
   h. $100 \times 0.048 = 4.8$
   i. $1000 \times 0.048 = 48$

2. a. $100 \times 5.439 = 543.9$
   b. $100 \times 4.03 = 403$
   c. $1000 \times 3.06 = 3060$
   d. $100 \times 30.54 = 3054$
   e. $30.73 \times 10 = 307.3$
   f. $93.103 \times 100 = 9310.3$

3.

a. $10^2 \times 0.007 = 0.7$ $10^3 \times 2.01 = 2{,}010$ $10^5 \times 4.1 = 410{,}000$	b. $10^5 \times 41.59 = 4{,}159{,}000$ $3.06 \times 10^4 = 30{,}600$ $0.046 \times 10^6 = 46{,}000$

4.

a. $0.4 \div 10 = 0.04$ $0.4 \div 100 = 0.004$ $4.4 \div 100 = 0.044$	b. $15.4 \div 100 = 0.154$ $21.03 \div 10 = 2.103$ $0.39 \div 10 = 0.039$	c. $5.6 \div 10 = 0.56$ $34.9 \div 100 = 0.349$ $230 \div 1000 = 0.23$

5.

a. $0.7 \div 10^2 = 0.007$ $45.3 \div 10^3 = 0.0453$ $568 \div 10^5 = 0.00568$	b. $2.1 \div 10^4 = 0.00021$ $4{,}500 \div 10^6 = 0.0045$ $9.13 \div 10^3 = 0.00913$

6.

a. $\frac{2}{100} = 0.02$ $\frac{2.1}{100} = 0.021$	b. $\frac{49}{1000} = 0.049$ $\frac{490}{1000} = 0.49$	c. $\frac{6}{10} = 0.6$ $\frac{6.5}{10} = 0.65$
d. $\frac{5}{10} = 0.5$ $\frac{5.04}{10} = 0.504$	e. $\frac{4.7}{10} = 0.47$ $\frac{4.7}{100} = 0.047$	f. $\frac{72}{100} = 0.72$ $\frac{72.9}{100} = 0.729$

7. One pound costs $72 ÷ 10 = $7.20.

8. 100 × $0.89 = $89.

9.

a. $\frac{239}{1000} = 0.239$	b. $\frac{35{,}403}{1000} = 35.403$	c. $\frac{67}{1000} = 0.067$
d. $\frac{263{,}000}{1000} = 263$	e. $\frac{3{,}890}{1000} = 3.89$	f. $\frac{1{,}692{,}400}{1000} = 1692.4$
g. $\frac{12{,}560{,}000}{1000} = 12{,}560$	h. $\frac{9}{1000} = 0.009$	i. $\frac{506{,}940}{1000} = 506.94$

Multiply and Divide by Powers of Ten, cont.

10.

a. $\frac{239}{100} = 2.39$	d. $\frac{89,803}{100} = 898.03$	g. $\frac{69}{10} = 6.9$
b. $\frac{239}{10} = 23.9$	e. $\frac{26,600}{100} = 266$	h. $\frac{69}{100} = 0.69$
c. $\frac{23,133}{100} = 231.33$	f. $\frac{3,402}{100} = 34.02$	i. $\frac{9}{10} = 0.9$

11. a. \$0.80 b. \$2.55 c. \$12.60

12. a. \$0.78 b. \$0.04 c. \$3.90

13. The new price is \$20.30. You can first find 1/10 of the price: \$29 ÷ 10 = \$2.90. Multiply that by 7 to get the new price (because when something is discounted by 3/10, it means 7/10 of the price is left). Alternatively, you can first calculate 3/10 of the old price, and subtract that from \$29.

14. a. The new price is \$100.80. We can simply find 8/10 of the old price: \$126 ÷ 10 = \$12.60, and 8 × \$12.60 = \$100.80. Alternatively, you can find 2/10 of the old price, and subtract that from \$126.
    b. The new price is \$42.75. We can find 5/100 of the old price: \$45 ÷ 100 = \$0.45, and 5 × \$0.45 = \$2.25. Then, we subtract that from \$45 to get \$42.75. Alternatively, you can find 95/100 of the old price.

15. The number is 100 × 0.03 = 3.

16. Model A: \$86.90 ÷ 10 × 7 = \$60.83
    Model B: \$75 ÷ 4 ×3 = \$56.25 Model B is cheaper.

17. *Why didn't 7 understand what 3.14 was talking about?* He didn't see his point.

4,200	1000		100	35.5	100	0.042		10		0.355	1000	1000
H	E		D	I	D	N	'	T		S	E	E

230	100	10		23	1000	4.2	4,200	42
H	I	S		P	O	I	N	T

Divide Decimals by Decimals 1, p. 177

Teaching Box:

You have learned:		
• ...how to **divide decimals by whole numbers,** using either mental math or long division.	Solve. 2.04 ÷ 2 = 1.02 0.24 ÷ 6 = 0.04 5.2 ÷ 10 = 0.52 5.2 ÷ 100 = 0.052	7) 17.22 = 2.46 -1 4 3 2 -2 8 4 2 -4 2 0
• ...how to **divide decimals by decimals mentally**, thinking of *how many times does it fit*:	Solve. 2.5 ÷ 0.5 = 5 0.021 ÷ 0.003 = 7	

1.

a. 60 ÷ 20 = 3 b. 6 ÷ 2 = 3 c. 0.6 ÷ 0.2 = 3 d. 0.06 ÷ 0.02 = 3	e. 350 ÷ 50 = 7 f. 35 ÷ 5 = 7 g. 3.5 ÷ 0.5 = 7 h. 0.35 ÷ 0.05 = 7	i. 2,000 ÷ 10 = 200 j. 200 ÷ 1 = 200 k. 20 ÷ 0.1 = 200 l. 2 ÷ 0.01 = 200

Divide Decimals by Decimals, cont.

2.

a. 5 ÷ 0.2 = 25 50 ÷ 2 = 25	b. 7 ÷ 0.35 = 20 700 ÷ 35 = 20	c. 36.9 ÷ 3 = 12.3 0.369 ÷ 0.03 = 12.3

3.

a. 0.445 ÷ 0.05	b. 2.394 ÷ 0.7
4.45 ÷ 0.5 44.5 ÷ 5	23.94 ÷ 7
8.9 5) 4 4.5 -4 0 4 5 - 4 5 0	3.4 2 7) 2 3.9 4 -2 1 2 9 - 2 8 1 4 - 1 4 0

4.

a. 0.832 ÷ 0.4 8.32 ÷ 4 = 2.08	b. 0.477 ÷ 0.09 4.77 ÷ 0.9 47.7 ÷ 9 = 5.3
c. 9.735 ÷ 0.003 97.35 ÷ 0.03 973.5 ÷ 0.3 9,735 ÷ 3 = 3,245	d. 1.764 ÷ 0.006 17.64 ÷ 0.06 176.4 ÷ 0.6 1,764 ÷ 6 = 294
e. 2.805 ÷ 0.11 28.05 ÷ 1.1 280.5 ÷ 11 = 25.5	f. 546.6 ÷ 1.2 5,466 ÷ 12 = 455.5

Divide Decimals by Decimals 2, p. 180

1.

a.	b.
a. 0.6 ÷ 0.02 = 30 6 ÷ 0.2 = 30 60 ÷ 2 = 30 600 ÷ 20 = 30	b. 0.48 ÷ 0.006 = 80 4.8 ÷ 0.06 = 80 48 ÷ 0.6 = 80 480 ÷ 6 = 80

2.

a.	b.
a. 44.7 ÷ 0.05 447 ÷ 0.5 4470 ÷ 5 = 894	b. 7.588 ÷ 0.007 75.88 ÷ 0.07 758.8 ÷ 0.7 7588 ÷ 7 = 1,084
8 9 4 5) 4 4 7 0 -4 0 4 7 - 4 5 2 0 - 2 0 0	1 0 8 4 7) 7 5 8 8 -7 0 5 - 0 5 8 - 5 6 2 8 - 2 8 0

3. a. 26 workbooks. The division is 8 ÷ 0.3, which gets transformed into 80 ÷ 3 = 26.666....
   b. There are about 8.1 km in 5 miles. The division is 5 ÷ 0.62, which is transformed into 50 ÷ 6.2 and then into 500 ÷ 62 ≈ 8.064.

4.

C 531.3 ÷ 0.11 = 4830
R 84.42 ÷ 0.09 = 938
A 143.696 ÷ 0.004 = 35924
M 44.615 ÷ 0.05 = 892.3
L 224.4 ÷ 0.24 = 935

H 394.4 ÷ 0.8 = 493
E 15.57 ÷ 0.45 = 34.6
T 8.725 ÷ 2.5 = 3.49
I 45.144 ÷ 1.9 = 23.76
N 2.802 ÷ 0.002 = 1401

T 19.992 ÷ 0.51 = 39.2
U 10.401 ÷ 0.3 = 34.67
B 38.142 ÷ 0.13 = 293.4
S 6.258 ÷ 0.07 = 89.4

What occurs once in a minute, twice in a moment and never in a thousand years? The letter "M".

3.49	493	34.6		935	34.6	39.2	39.2	34.6	938			892.3	
T	H	E		L	E	T	T	E	R		"	M	"

What is so fragile that even saying its name can break it? Silence.

89.4	23.76	935	34.6	1401	4830	34.6
S	I	L	E	N	C	E

What goes up a chimney down, but won't go down a chimney up? Umbrella.

34.67	892.3	293.4	938	34.6	935	935	35924
U	M	B	R	E	L	L	A

5. a. 1.66 ÷ 0.03 = 55.3. Transform 1.66 ÷ 0.03 into 16.6 ÷ 0.3 and then into 166 ÷ 3. Remember to use 166.00 in the long division, so you can get the answer to 2 decimal digits, and then round it to 1 decimal digit.
   b. 28 ÷ 0.3 = 93.33. Transform 28 ÷ 0.3 into 280 ÷ 3. Remember to use 280.000 in the long division, so you can get the answer to 3 decimal digits, and then round it to 2 decimal digits.
   c. 0.39 ÷ 0.007 = 55.714. Transform 0.39 ÷ 0.007 into 3.9 ÷ 0.07, then into 39 ÷ 0.7, and then into 390 ÷ 7. Use 39.0000 in the long division, so you can get the answer to 4 decimal digits, and then round it to 3 decimal digits.
   d. 45 ÷ 21 = 2.14 Remember to use 45.000 in the long division so you can get the answer to 3 decimal digits, and can then round it to 2 decimal digits.

6. a. 200.9 ÷ 3 = 66.96 R0.2
   b. 430 ÷ 9 = 47.7 R0.7
   c. 52 ÷ 6 = 8.666 R0.004
   d. 45 ÷ 21 = 2.14 R0.06

Decimals in Measuring Units and More, p. 184

1.

a. 0.8 m = 80 cm 1.3 m = 130 cm 8.27 m = 827 cm	b. 56 cm = 0.56 m 3 cm = 0.03 m 382 cm = 3.82 m	c. 0.31 m = 31 cm 4.6 m = 460 cm 16.08 m = 1608 cm

2.

a. 0.5 km = 500 m 0.7 km = 700 m 4.5 km = 4,500 m	b. 2,400 m = 2.4 km 12,680 m = 12.68 km 540 m = 0.54 km	c. 2.001 km = 2,001 m 0.319 km = 319 m 0.04 km = 40 m

3. Jake is 1.88 m − 0.16 m = 1.72 m or 172 cm tall.

4. 4 × 550 m = 2,200 m = 2.2 km

5. (2,400 + 350) × 2 = 5,500 m or 5.5 km

6.

a. 0.7 L = 700 ml 12.6 L = 12,600 ml 0.06 L = 60 ml	b. 3,900 ml = 3.9 L 2,080 ml = 2.08 L 212 ml = 0.212 L	c. 0.009 L = 9 ml 1.35 L = 1,350 ml 1.585 L = 1,585 ml

7. The difference is 520 ml − 470 ml = 50 ml.

Teaching box:

- 0.1 kg is 100 g
- 0.01 kg is 10 g
- 0.001 kg is 1 g

8.

a. 0.3 kg = 300 g 2.6 kg = 2,600 g 0.05 kg = 50 g	b. 20 g = 0.02 kg 800 g = 0.8 kg 6,030 g = 6.03 kg	c. 1.1 kg = 1,100 g 0.152 kg = 152 g 2.093 kg = 2,093 g

9. The apple weighed 180 g. 3/4 of 180 g is 180 g ÷ 4 × 3 = 135 g.

10. 30 × 450 g = 13,500 g = 13.5 kg

11. a. 0.2 mi = 1,056 ft b. 1.35 mi = 7,128 ft
c. 2.7 mi = 4,752 yd d. 0.72 mi = 1,267 yd

12. 0.7 mi + 0.65 mi + 0.5 mi × 3 = 2.85 mi = 15,048 ft

13.

a. 0.65 mi = 3,432 ft 1.34 mi = 7,075 ft	b. 0.9 mi = 1,584 yd 5.413 mi = 9,527 yd	c. 5.428 mi = 28,660 ft 2.75 mi = 4,840 yd

14. You need 55 McKinley mountains need stacked on top of each other. Convert either Mt. McKinley's height into miles or the distance 211.3 miles into feet before calculating. For example, Mt. McKinley is 20,320 ÷ 5,280 = 3.8484 miles tall. Then, divide to find how many of them are needed to reach the height of the International Space Station:
211.3 mi ÷ 3.8484 = 54.905 ≈ 55.

Decimals in Measuring Units and More, cont.

15.

a. \$0.7 M = \$700,000 \$2.5 M = \$2,500,000	b. \$0.04 M = \$40,000 \$0.39 M = \$390,000	c. \$10.9 M = \$10,900,000 \$2.78 M = \$2,780,000

16. They spent \$2.85 M − \$0.35 M = \$2.5 M = \$2,500,000.

17. The men's record advanced 8.95 m − 8.21 m = 0.74 m = 74 cm. The women's record advanced 7.52 m − 6.40 m = 1.12 m = 112 cm. The women's record has advanced 38 cm more than the men's.

18. (900 cm − 2 × 90 cm) ÷ 55 cm = 13.0909. You can place 13 chairs in the row.

Rounding and Estimating, p. 188

1.

a. 8.19 m ≈ 8 m; rounding error = 0.19 m	b. 362 cm ≈ 4 m; rounding error = 0.38 m	c. 417 cm ≈ 4 m; rounding error = 0.17 m
d. 1 m 54 cm ≈ 2 m; rounding error = 0.46 m	e. 14.208 m ≈ 14 m; rounding error = 0.208 m	f. 8 m 9 cm ≈ 8 m; rounding error = 0.09 m

2.

a. 602 m ≈ 1 km; rounding error = 0.398 km	b. 10.189 km ≈ 10 km; rounding error = 0.189 km	c. 8.057 km ≈ 8 km; rounding error = 0.057 km
d. 2,643 m ≈ 3 km; rounding error = 0.357 km	e. 6 km 55 m ≈ 6 km; rounding error = 0.055 km	f. 3,288 m ≈ 3 km; rounding error = 0.288 km

3. a. 1.5 m − 0.67 m = 0.83 m ≈ 0.8 m
   b. 6.08 m + 0.45 m + 1.2 m = 7.73 m ≈ 7.7 m
   c. 1.08 m + 2.55 m = 3.63 m ≈ 3.6 m

4. a. 2,100 m − 293 m = 1,807 m ≈ 1,810 m
   b. 6,070 m + 452 m = 6,522 m ≈ 6,520 m
   c. 2,075 m + 3,800 m = 5,875 m ≈ 5,880 m

5.

a. Estimation: 4 ÷ 0.80 = 5 books Exact answer: 4 ÷ 0.78 = 5.13, so 5 books will fit	b. Change to grams. Estimation: 400 g ÷ 25 g = 16 papers Exact answer: 400 g ÷ 24 g = 16.6, so you can mail 16 papers.

6. Answers vary. For example: 18 × 2.2 cm ≈ 20 × 2 cm = 40 cm. It is better to round both factors, one up to 20, and the other down to 2 cm, than to only round 2.2 cm to 2 cm and leave the 18 unrounded. You can see that by checking the exact answer: it is 39.6 cm, which is really close to our estimate of 40 cm.

7. Answers vary. Notice we need to change 146 cm into meters. For example: 2 ½ × 1.46 m ≈ 3 × 1.5 m = 4.5 cm. Or, 2 ½ × 1.46 m ≈ 3 × 1.4 m = 4.2 m. Or, 2 ½ × 1.46 m ≈ 2 ½ × 1.5 m = 3 m + 0.75 m = 3.75 m. The last estimation is the most accurate.

8. He jogs 3 × 1,425 m = 4,275 meters, which is about 4,300 meters.

9. a. \$50 – (10 × \$2.46)

   b.

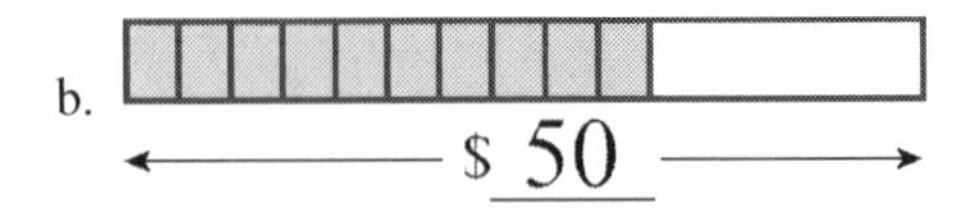

   c. He has left about \$50 – (10 × \$2.50) = \$25.
   d. He has left \$50 – (10 × \$2.46) = \$25.40.

The Metric System, p. 190

1.

a. 2 cm = 2/100 m = 0.02 m 6 dm = 6/10 m = 0.6 m 8 mm = 8/1000 m = 0.008 m	b. 3 dam = 30 m 9 km = 9,000 m 2 hm = 200 m	c. 6 mm = 0.006 m 20 cm = 0.20 m 8 dm = 0.8 m

2.

a. 2 ml = 2/1000 L = 0.002 L 6 cl = 6/100 L = 0.06 L 8 dg = 8/10 g = 0.8 g	b. 7 dl = 0.7 L 6 mg = 0.006 g 8 dl = 0.8 L	c. 3 dag = 30 g 8 kg = 8,000 g 2 hl = 200 L

3.

a. 75.4 m

		7	5.	4		
km	hm	dam	m	dm	cm	mm

b. 843 mm

				8	4	3
km	hm	dam	m	dm	cm	mm

c. 4.6 km

4.	6					
km	hm	dam	m	dm	cm	mm

d. 35.49 dam

	3	5.	4	9		
km	hm	dam	m	dm	cm	mm

4.

	to m	to dm	to cm	to mm
a. 75.4 m	75.4	754	7,540	75,400
b. 843 mm	0.843	8.43	84.3	843

5.

	to hm	to dam	to m	to dm
a. 4.6 km	46	460	4,600	46,000
b. 35.49 dam	3.549	35.49	354.9	3,549

6. a. hectograms b. centigrams c. decigrams

7. a. 4,500 dl b. 450.0 L (or 450 L) c. 45.00 dal (or 45 dal) d. 4.500 hl (or 4.5 hl)

8.

	a. 5,000 mm	b. 380 cm	c. 6.5 dm
meters	5 m	3.8 m	0.65 m
decimeters	50 dm	38 dm	6.5 dm
centimeters	500 cm	380 cm	65 cm
millimeters	5,000 mm	3,800 mm	650 mm

9. It will last 10 days. 200 ml is equal to 20 cl.

10. a. Hannah 151 cm; Erica 136 cm b. Hannah 175 cm; Erica 160 cm c. Hannah 165 cm; Erica 150 cm

11. a. 200 × 14 dg = 2,800 dg = 280 g. b. four boxes

Converting Between Customary Units of Measurement, p. 193

1.

a. 6 ft = 72in 7 ft 5 in. = 89 in	b. 25 in = 2 ft 1 in 45 in = 3 ft 9 in	c. 13 ft 7 in = 163 in 71 in. = 5 ft 11 in

2.

a. 2 lb 8 oz = 40 oz 45 oz = 2 lb 13 oz	b. 8 lb = 128 oz 56 oz = 3 lb 8 oz	c. 43 oz = 2 lb 11 oz 90 oz = 5 lb 10 oz

3.

a. 3 C = 24 oz 55 oz = 6 C 7 oz	b. 4 C = 2 pt 3 pt = 6 C	c. 7 gal = 28 qt 45 qt = 11 gal 1 qt

4. a. Multiply: 11 yd × 3 × 12 = 396 in.
   b. Divide: 711 ÷ 12 = 59 R3, so 711 in. = 59 ft 3 in.
   c. Divide: 982 ÷ 12 = 81 R10, so 982 in. = 81 ft 10 in. Then divide 81 ft by 3 to get 27 yd. So, 982 in. = 27 yd 0 ft 10 in.
   d. First, divide by eight: 254 ÷ 8 = 31 R6, so 254 oz = 31 C 6 oz.
   Next, convert the 31 cups into quarts and cups: 31 C ÷ 4 = 7 qt 3 C. We get 254 oz = 7 qt 3 C 6 oz.
   Lastly, write 7 qt as 1 gal 3 qt., and get 254 oz = 1 gal 3 qt 3 C 6 oz.

5.

a. 7.4 mi = 39,072 ft 16,000 ft = 3.03 mi	b. 1,500 ft = 500 yd 7,500 yd = 4.26 mi	1,760: mile mi → yard yd 3: yard yd → foot ft 12: foot ft → inch in 1 mile = 5,280 feet
c. 900 ft = 0.17 mi 2.56 mi = 4505.6 yd	d. 12.54 mi = 66211.2 ft 82,000 ft = 15.53 mi	

6.

a. 15.2 lb = 243.2 oz 655 oz = 40.94 lb	b. 4.78 T = 9,560 lb 7,550 lb = 3.78 T	c. 78 oz = 4.88 lb 0.702 T = 1,404 lb

7.

F 0.6 mi = 3,168 ft	**G** 7 C = 56 oz	**I** 14,256 ft = 2.7 mi
A 5,632 yd = 3.2 mi	**R** 6,200 lb = 3.1 T	**W** 6 ft 7 in = 79 in
O 10 qt = 40 C	**S** 3 lb 5 oz = 53 oz	**L** 732 in = 61 ft
H 2 lb 11 oz = 43 oz	**E** 5 ft 2 in = 62 in	**D** 42 in = 3.5 ft
L 1.3 mi = 2,288 yd	**O** 40 oz = 2.5 lb	**P** 3 gal = 24 pt
		A 0.75 mi = 3,960 ft

What did one potato chip say to the other?

53	43	3960	61	2288		79	62		56	40
S	H	A	L	L		W	E		G	O

3168	2.5	3.1		3.2		3.5	2.7	24	
F	O	R		A		D	I	P	?

8. a. Two gallons. There are 16 cups in 1 gallon. 2 × 16 C = 32 C; you will need two *whole* gallons to have 30 cups.
   b. She made 72 jars of applesauce. Once gallon is 4 quarts, so in quarts, mom got 2 × 9 × 4 qt = 72 qt.
   c. You can cut out fourteen 8-inch pieces. 9 3/4 ft in inches is 9.75 × 12 in. = 117 in. Then, divide that by 8 in: 117 in. ÷ 8 in = 14 R5, so you can cut fourteen 8-inch pieces and you'll have 5 inches left over.
   d. One ounce costs $1.20 ÷ 4 = $0.30,so 5 ounces cost 5 × $0.30 = 1.50.
   e. It weighs 16 lb 12 oz. In ounces, the box weighs 20 × 13 oz + 8 oz = 268 oz. To convert that to pounds, divide by 16: 268 oz ÷ 16 oz = 16 R12. The box of shampoo weighs 16 lb 12 oz.
   f. Either 56 C 2 oz, or 14 qt 2 oz, or 3 gal 2 qt 2 oz. In a month, Mark drinks 3 × 5 oz × 30 = 450 oz. To convert that to cups and ounces, divide by eight: 450 ÷ 8 = 56 R2, so 450 oz = 56 C 2 oz. To further convert the 56 cups into quarts, divide by four: 56 ÷ 4 = 14, so 56 C = 14 qt. And, 14 qt can be written as 3 gal 2 qt.
   g. On average, she lost 2.86 ounces per day. She lost 5 × 16 = 80 oz in 28 days. 80 ÷ 28 ≈ 2.857.

Number Rule Puzzles, p. 197

1. The rule: A − B = C or C + B = A or B + C = A.

A	22	9,000	0.5	4.5	1.65	6.7	0.43	0.49	6.69
B	15	2,301	0.05	4.2	0.05	0.09	0.38	0.48	3.19
C	7	6,699	0.45	0.3	1.6	6.61	0.05	0.01	3.5

2. The rule: A ÷ 7 = B or B = A ÷ 7 or A = 7B.

A	5.6	112	0.49	0.63	8456	0.42	3.43	34.3	0.063
B	0.8	16	0.07	0.09	1208	0.06	0.49	4.9	0.009

3. The rule: A + 1 = B or B = A + 1 or A = B − 1.

A	0.42	2,001	0.083	0.9	0.04	9.023	179	1.032	1.05
B	1.42	2,002	1.083	1.9	1.04	10.023	180	2.032	2.05

4. The rule: B = 3 × A or A = B ÷ 3.

A	1	0	0.3	1.5	0.6	1.6	0.5	0.2	0.04
B	3	0	0.9	4.5	1.8	4.8	1.5	0.6	0.12

Problem Solving, p. 198

1. a. The piece was originally 6.03 m long. (4.69 m ÷ 7 × 9 = 6.03 m)

   b. The piece cut off was 1.34 m.

?

4.69 m

2. a. A bouquet of roses costs \$20.80. (\$15.60 ÷ 3 × 4 = \$20.80)

   b. 2 × \$15.60 + 3 × \$20.80 = \$93.60

?

\$15.60

3. The new price is \$35.60. You can calculate it in several different ways.
   For example, you can find 8/10 of the original price: 0.8 × \$44.50 = \$35.60.
   Or, do the same using this calculation: \$44.50 ÷ 10 × 8 = \$35.60.
   Or, first calculate 2/10 of the price, and then subtract that from the original price:
   \$44.50 ÷ 10 × 2 = \$8.90; \$44.50 – \$8.90 = \$35.60.

4. a. The sandals are \$8.75. (0.7 × \$12.50 = \$8.75) b. The tennis shoes cost \$18.13 (0.7 × \$25.90 = \$18.13)
   c. \$26.88

5. The cheaper rake costs (\$22.70 – \$5.60) ÷ 2 = \$8.55.

6. The lighter weight weighs (5.66 kg – 1.5 kg) ÷ 2 = 2.08 kg.
   The heavier weight weighs 2.08 kg + 1.5 kg = 3.58 kg.

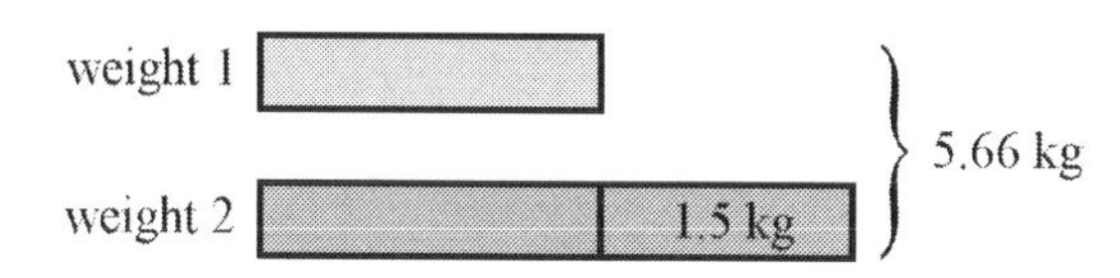

7. One cheaper spade costs (\$6.90 – \$1.50) ÷ 2 = \$2.70.
   Three of them cost 3 × \$2.70 = \$8.10.

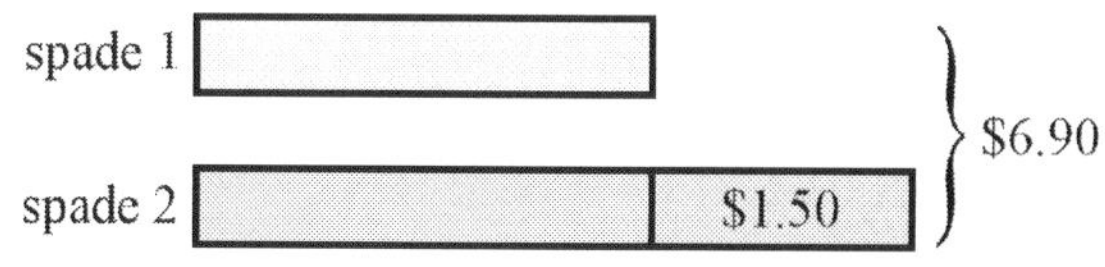

8. Lily paid \$3.90 and Brenda paid \$7.80. (Lily paid 1/3 of the total cost and Brenda paid 2/3 of it, so you will find Lily's share by dividing by 3.)

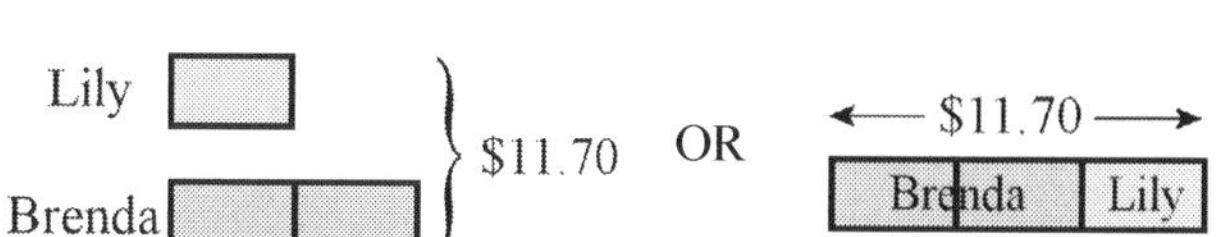

Mystery number: 16 and 17, because 16 × 17 – 12 = 260.

Mixed Review, p. 202

1. a. 1,070 b. 2,515 c. 901

2. a. y = 62,103 (subtract 21,390 from 83,493) b. s = 317 (divide 6,340 by 20)

3. a. 10 b. 30 c. 58 d. 146

4. a. 15,000,600,024 b. 42,080,017

5. a. 1, 2, 3, 6, 7, 14, 21, 42 b. 1, 2, 4, 8, 16, 32, 64

6. a. Twenty song downloads would cost \$2.20. \$5.50 ÷ 50 × 20 = \$2.20.

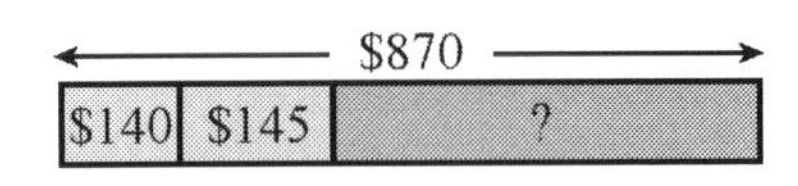

b. Mr. Doe had \$585 left. \$870 – (\$870 ÷ 6) – \$140 = \$585

c. Henry owns 4 × 450 = 1,800 stamps.

d. The cheaper hammer cost (\$64 – \$28) ÷ 2 = \$18.

Hammer 1 ?
Hammer 2 $28
$64

Mystery number: 264

Review, p. 204

1.

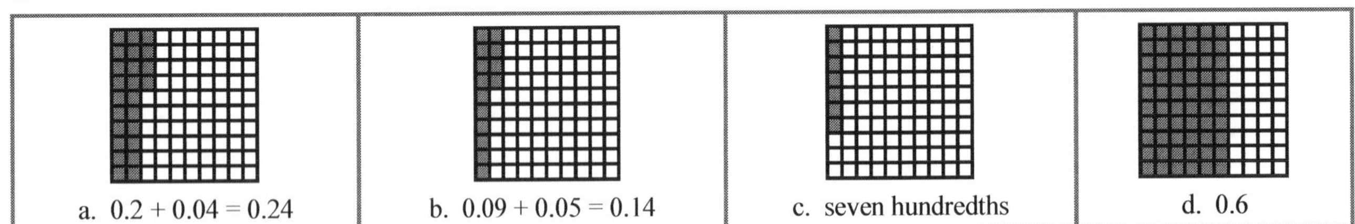

a. 0.2 + 0.04 = 0.24 b. 0.09 + 0.05 = 0.14 c. seven hundredths d. 0.6

2. a. 0.495 = 4 × (1/10) + 9 × (1/100) + 5 × (1/1000)

b. 2.67 = 2 × 1 + 6 × (1/10) + 7 × (1/100)

3. a. 0.042 b. 0.047 c. 0.05 d. 0.055 e. 0.062

4. a. > b. > c. < d. > e. <

5. a. 0.03 b. 0.048 c. 1.209 d. 3.39

6. a. $1\frac{3}{10}$ b. $2\frac{15}{100}$ c. $\frac{8}{1000}$ d. $\frac{38}{1000}$

7.

rounded to...	nearest one	nearest tenth	nearest hundredth
4.608	5	4.6	4.61
3.109	3	3.1	3.11
2.299	2	2.3	2.30
0.048	0	0.0	0.05

8.

a. 0.3 + 0.005 = 0.305 0.03 + 0.5 = 0.53	b. 0.9 – 0.7 = 0.2 0.9 – 0.07 = 0.83
c. 0.008 + 0.9 + 5 = 5.908 0.9 + 0.8 + 0.17 = 1.87	d. 2.5 – 1.02 = 1.48 7.8 – 0.9 – 0.04 = 6.86

9. a. 0.21 + 0.79 = 1 b. 0.004 + 0.996 = 1 c. 4.391 + 0.609 = 5

10. a. 3.944 b. 0.099

Review, cont.

11. a. Each child should have about \$3.00 left. b. $25 \div 5 - \$2.05$ c. Each child has \$2.95 left.

12.

a. $0.4 \times 8 = 3.2$ b. $6 \times 0.009 = 0.054$	c. $20 \times 0.5 = 10$ d. $100 \times 0.3 = 30$	e. $0.9 \times 0.2 = 0.18$ f. $0.06 \times 0.3 = 0.018$

13.

a. $0.35 \div 5 = 0.07$ b. $4.5 \div 9 = 0.5$	c. $0.4 \div 10 = 0.04$ d. $5 \div 100 = 0.05$	e. $0.38 \div 10 = 0.038$ f. $7 \div 1000 = 0.007$

14.

a. $0.8 \times 0.5 = 0.40$ b. $8 \times 0.008 = 0.064$	c. $7 \times 0.5 = 3.5$ d. $0.6 \times 0.04 = 0.024$	e. $0.9 \times 8 = 7.2$ f. $9 \times 0.09 = 0.81$

15.

a. $0.07 \times 10^2 = 7$ $10^5 \times 1.08 = 108{,}000$	b. $3{,}300 \div 10^4 = 0.33$ $239.8 \div 10^3 = 0.2398$

16. a. 0.7×5 kg = 3.5 kg b. 0.06×1.2 m = 0.072 m c. 0.35×2 L = 0.7 L

17. a. 1.817 b. 0.355 c. 0.85

18. Answers will vary. See some example answers below:
The answer is less than 0.7 because you multiply 0.7 by a number that is less than one.
The answer has to be less than 0.7 because multiplying by 0.4 means you are taking a fractional part of 0.7.
Multiplying by 0.4 means taking 4/10 part of 0.7, and 4/10 is less than 1, so the answer is less than 0.7

19. The discounted price is \$100.80. For example, you can calculate $0.8 \times \$126 = \$100.80$.

20. a. 382. The division $152.8 \div 0.4$ is converted into $1{,}528 \div 4 = 382$.
b. 34.7. The division $2.776 \div 0.08$ is converted into $27.76 \div 0.8$ and then into $277.6 \div 8 = 34.7$.
c. 163.64. The division $180 \div 1.1$ is converted into $1800 \div 11 \approx 163.636$.
d. 0.29. The division $2.000 \div 7 = 0.285$ R0.005.

21.

a. 0.9 m = 90 cm 45 cm = 0.45 m 1.5 km = 1,500 m	b. 0.6 L = 600 ml 5,694 ml = 5.694 L 0.09 L = 90 ml	c. 2.2 kg = 2,200 g 390 g = 0.390 kg 0.02 kg = 20 g

22.

a. 6 ft 11 in. = 83 in. 3 lb 11 oz = 59 oz 3 C = 24 oz	b. 2 gal = 32 C 5 qt = 10 pt 54 oz = 6 C 6 oz	c. 78 oz = 4 lb 14 oz 39 in = 3 ft 3 in 102 in = 8 ft 6 in

23.

a. 2.65 mi = 13,992 ft 10.9 mi = 19,184 yd	b. 3,800 ft = 0.72 mi 3,500 yd = 1.99 mi	c. 4.54 lb = 72.64 oz 10.2 ft = 122.4 in

24. a. Each box weighs 5.2 kg or 5 1/5 kg. 26.0 kg $\div$ 5 = 5.2 kg or 26 kg $\div$ 5 = 5 1/5 kg
b. They cost \$15.60 per box. $5.2 \times 3 = 15.6$.

25. Edward earns \$446.50. Multiply $38 \times \$11.75 = \$446.50$ to get that. He pays in taxes $\$446.50 \div 5 = \$89.30$.
He takes home \$357.20 after taxes.

26. The smaller pitcher holds 1.55 liters and the larger holds 2.1 liters. (3.65 L − 0.55 L) $\div$ 2 = 3.1 L $\div$ 2 = 1.55 L.

Math Mammoth Grade 5-B
Answer Key

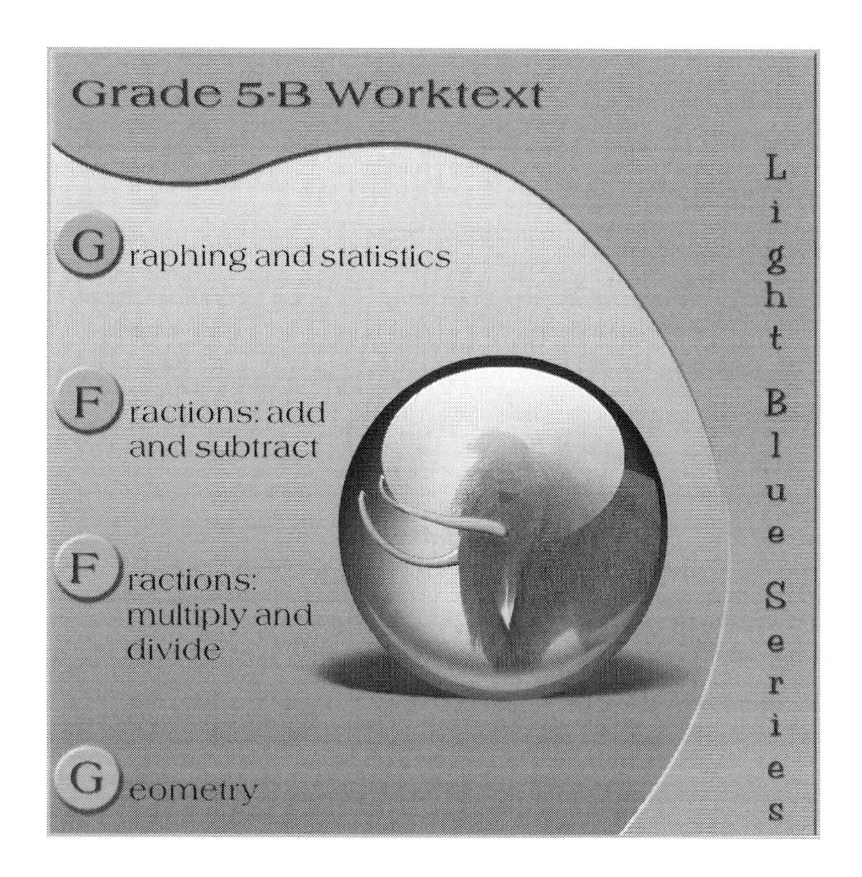

By Maria Miller

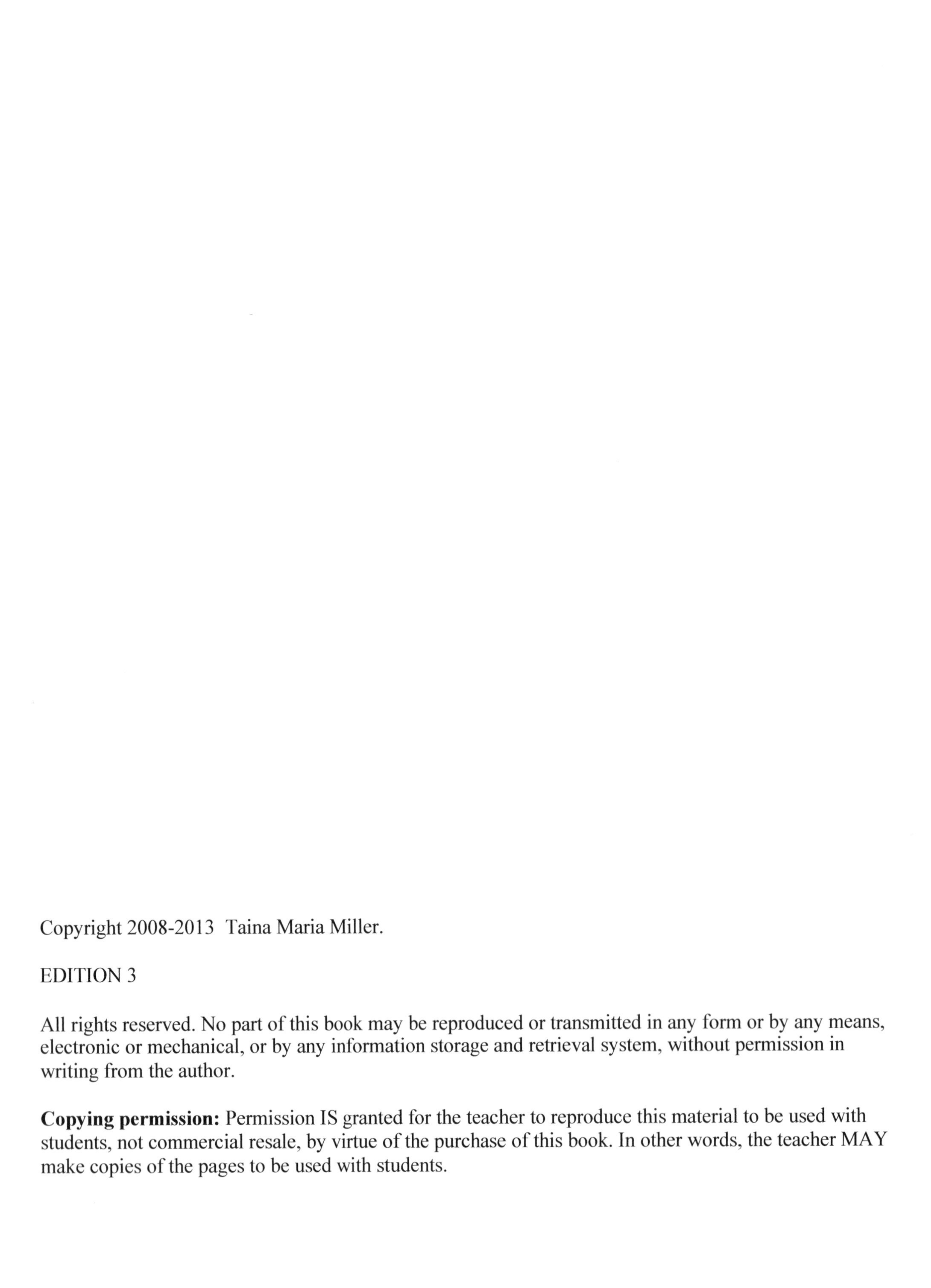

EDITION 3

Math Mammoth Grade 5-B Answer Key

Contents

Chapter 5: Graphing and Statistics

Coordinate Grid, p. 10

1. A (1, 2) B (3, 4) C (2, 9) D (6, 5)
   E (8, 3) F (8, 8) G (10, 9) H (10, 1)

2.

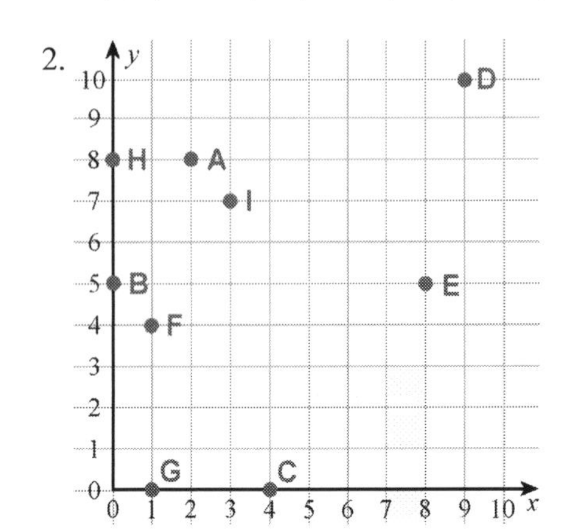

3. A house:

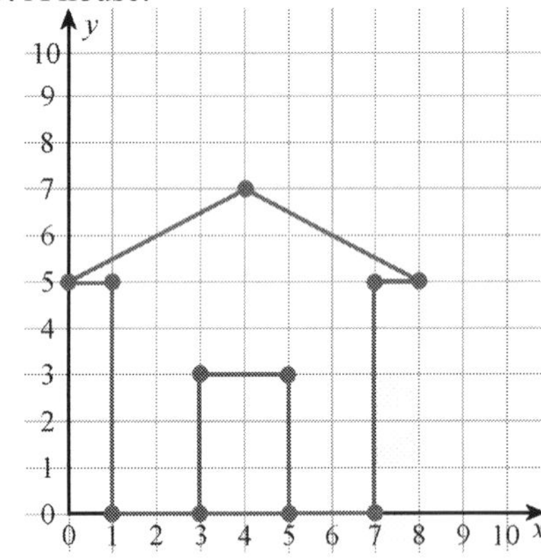

4. a.

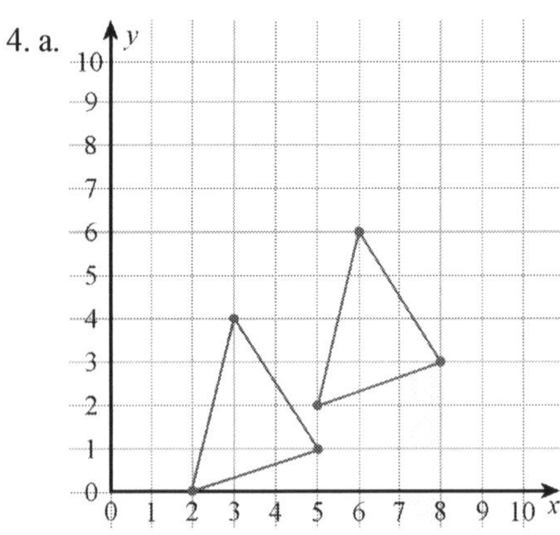

b. A (5, 2) B (8, 3) C (6, 6)

5. a. The line segment was moved three units down and two to the left.
   The original coordinates are A(8,6) B(10, 8) and C(9, 5).
   The coordinates of the moved triangle A′B′C′ are
   A′ (6, 3), B′(8, 5), and C′(7, 2).

   b. The point moves two units up and one unit to the left.
   The coordinates of the moved triangle are
   A(7,8), B(9, 10) and C(8, 7).

Number Patterns in the Coordinate Grid, p. 13

1. a.

b.

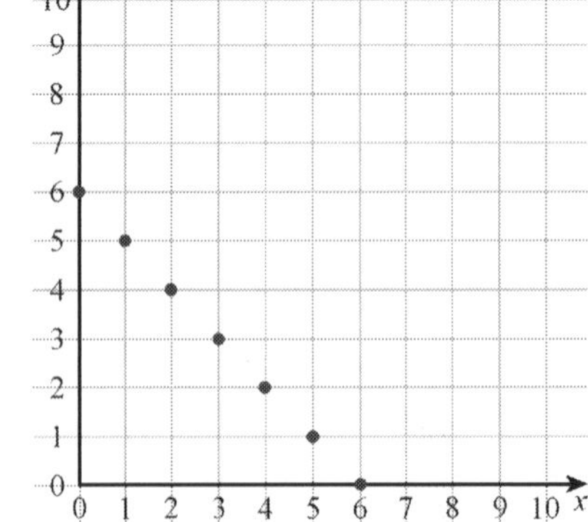

2. a.

x	0	2	4	6	8	10
y	0	1	2	3	4	5

b.

x	2	3	4	5	6	7
y	10	8	6	4	2	0

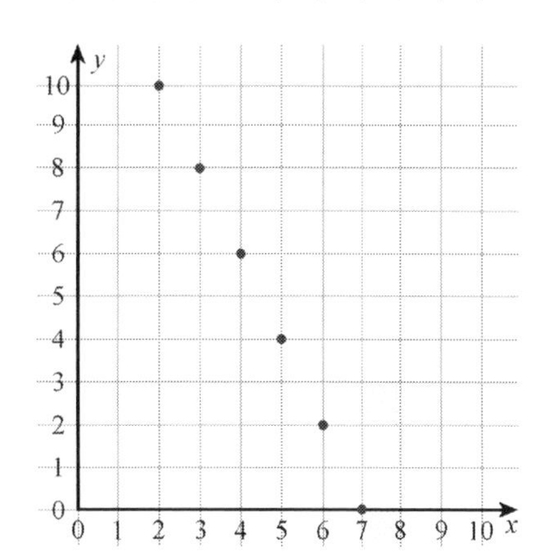

c. Answers vary. Check student's answers.

3.

x	0	10	20	30	40	50	60	70	80	90	100	110
y	2	3	4	5	6	7	8	9	10	11	12	13

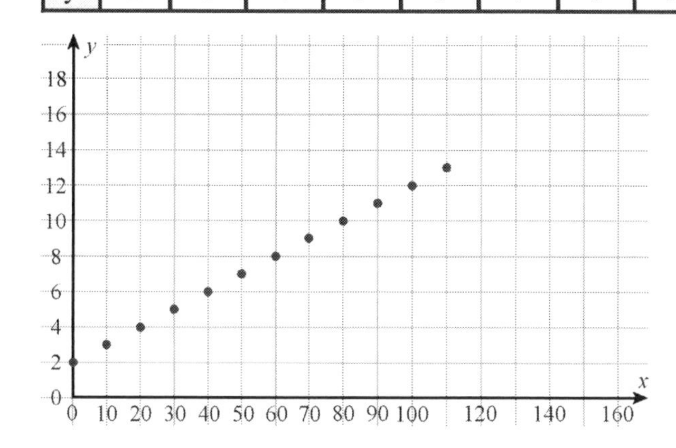

4. a. The rule is $y = x - 1$.

x	1	2	3	4	5	6
y	0	1	2	3	4	5

b. The rule is $y = 5x$.

x	1	2	3	4	5	6	7	8	9	10
y	5	10	15	20	25	30	35	40	45	50

5. a. (0, 5), (1, 4), (2, 3), (3, 2), (4, 1), and (5, 0)
b. The rule is $y = 5 - x$. Or, you can also write it as $x + y = 5$. Both are correct.

x	0	1	2	3	4	5
y	5	4	3	2	1	0

6. The answers will vary. Check the students' answers.

More Number Patterns in the Coordinate Grid, p. 17

1. **The rule for x-values:** start at 0, and add 1 each time.
   The rule for y-values: start at 1, and add 2 each time.

x	0	1	2	3	4	5	6
y	1	3	5	7	9	11	13

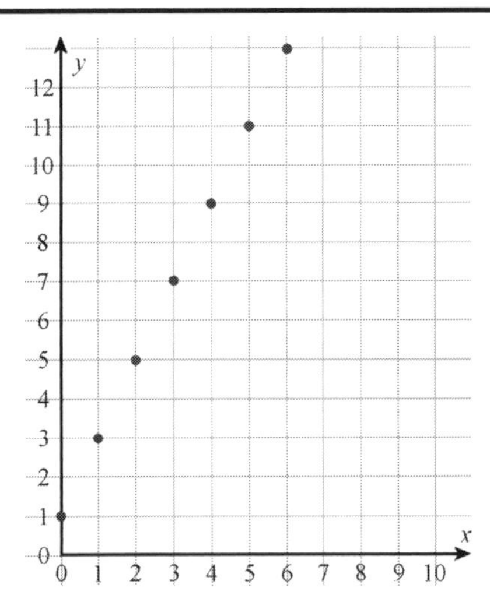

2. **The rule for x-values:** start at 10, and subtract 1 each time.
   The rule for y-values: start at 1, and add 2 each time.

x	10	9	8	7	6	5	4
y	1	3	5	7	9	11	13

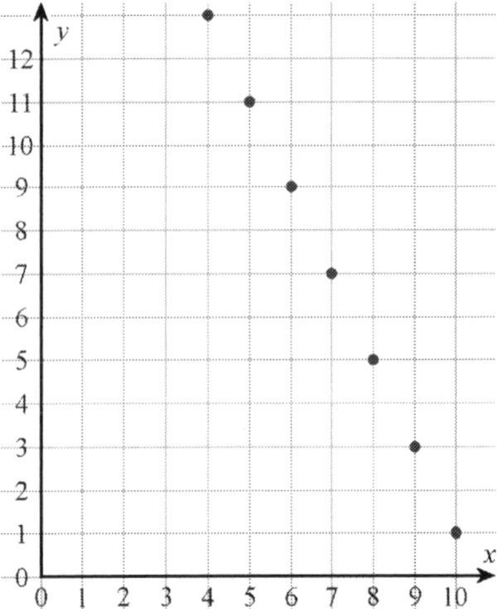

3. **The rule for x-values:** start at 1, and add 1 each time.
   The rule for y-values: start at 5, and subtract ½ each time.

x	1	2	3	4	5	6	7	8
y	5	4 ½	4	3 ½	3	2 ½	2	1 ½

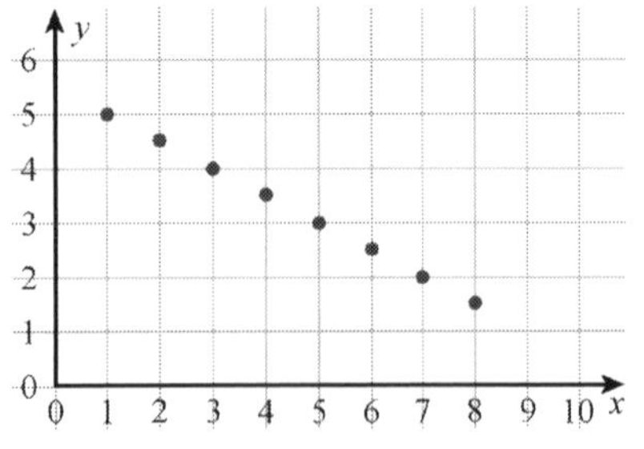

4. **The rule for x-values:** start at 8, and subtract ½ each time.
   The rule for y-values: start at 0, and add 1 each time.

x	8	7 ½	7	6 ½	6	5 ½	5	4 ½
y	0	1	2	3	4	5	6	7

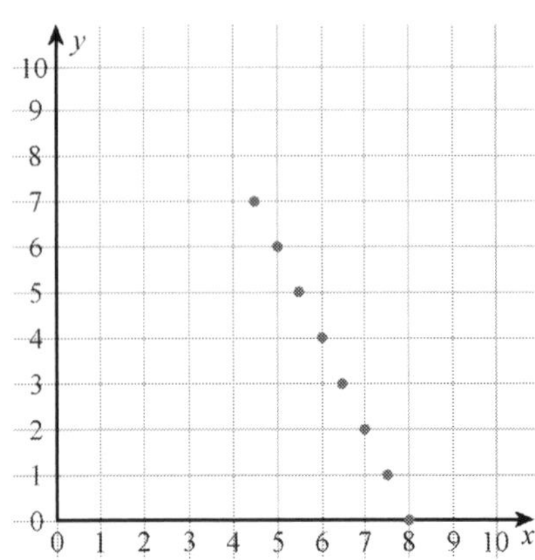

5. Answers vary.

6. Answers vary.

More Number Patterns in the Coordinate Grid, cont.

7. **The rule for *x*-values:** start at 0, and add 2 each time.
   The rule for *y*-values: start at 0, and add 1 each time.

x	0	2	4	6	8	10
y	0	1	2	3	4	5

The rule is: y is half of x. In other words, $y = x/2$.
Or, the other way around we can say x is double y, or $x = 2y$.

Explanations vary. For example:
Since x-values skip-count by 2s, while y-values only skip-count by ones, the x-values increase double as fast as the y-values. Thus, x ends up being double y. Or, skip-counting by 2 forms the multiplication table of 2, and that is why x is 2 times y.

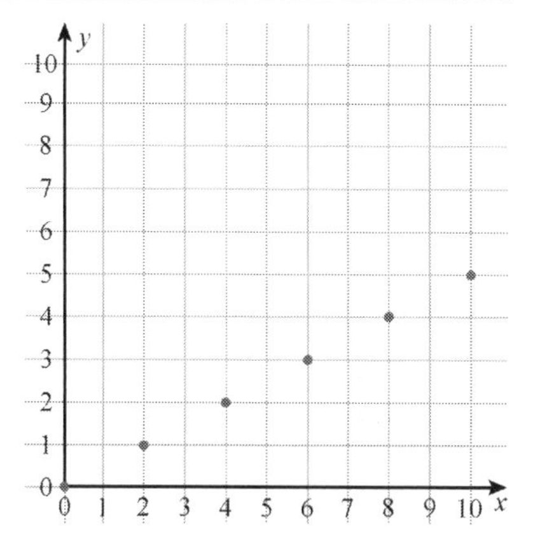

8. **The rule for x-values:** start at 0, and add ½ each time.
   The rule for y-values: start at 0, and add 1 each time.

x	0	½	1	1 ½	2	2½	3	3½
y	0	1	2	3	4	5	6	7

The rule is: y is double x. In other words, $y = 2x$.
Or, the other way around we can say x is half of y, or $x = y/2$.

Explanations vary. For example:
Since y-values skip-count by ones, while x-values only skip-count by halves, the y-values increase double as fast as the x-values. Thus, y ends up being double x.

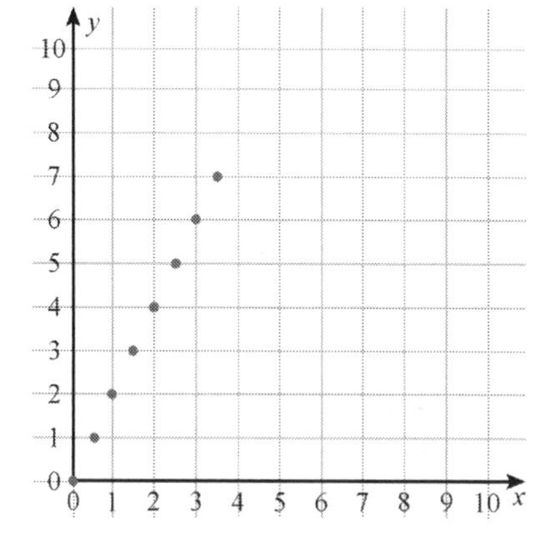

9.

x	0	10	20	30	40	50	60	70	80	90	100	110
y	0	1	2	3	4	5	6	7	8	9	10	11

The rule is: x is 10 times y, or $x = 10y$.

Explanations vary. For example: Skip-counting by 10s creates the multiplication table of 10, and that is why x ends up being 10 times y.

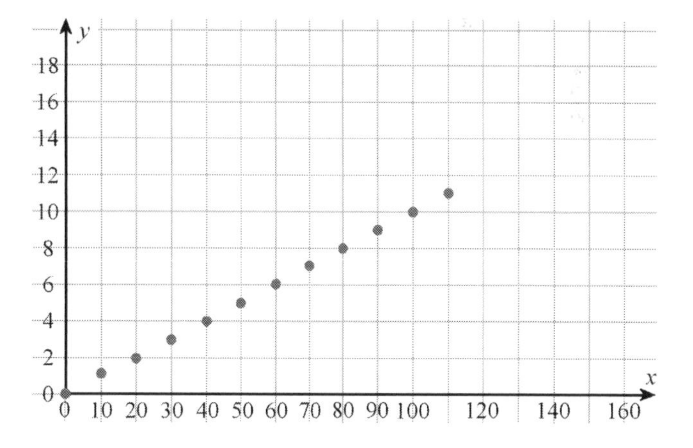

More Number Patterns in the Coordinate Grid, cont.

10.

x	0	2	4	6	8	10	12	14	16	18	20
y	0	½	1	1 ½	2	2 ½	3	3 ½	4	4 ½	5

The rule is: x is four times y, or $x = 4y$. Or, the other way around we can say y is one-fourth of x, or $y = x/4$.

Explanations vary. For example: Since x-values skip-count by 2s, while y-values skip-count by halves, the x-values increase four times as fast as the y-values. Thus, x ends up being four times y.

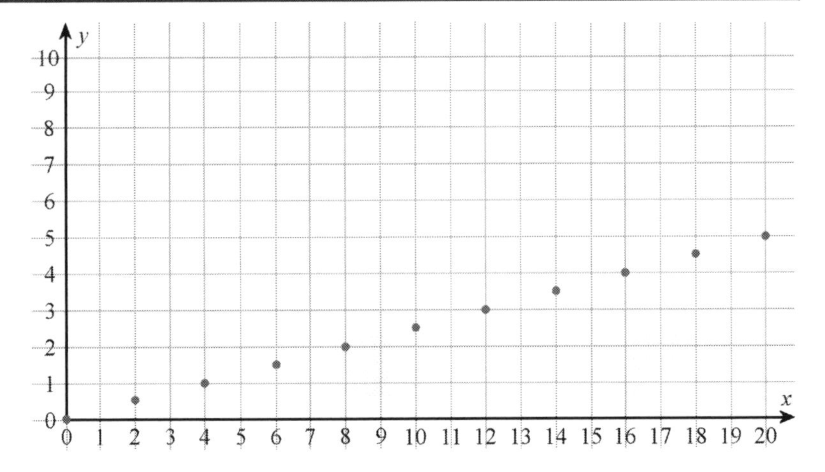

Line Graphs, p. 21

1. a, b, c:

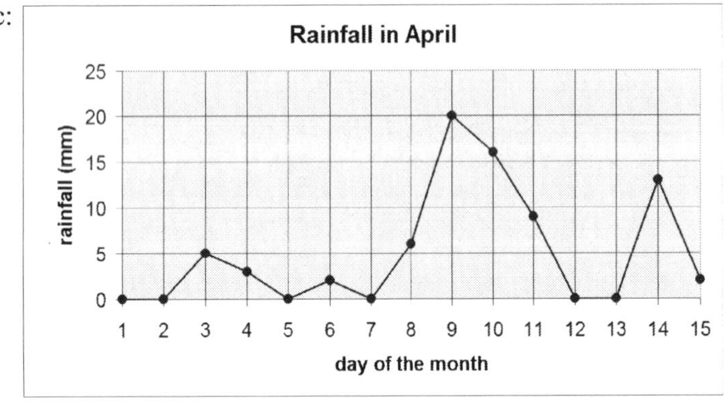

d. 6 days.

2. a.

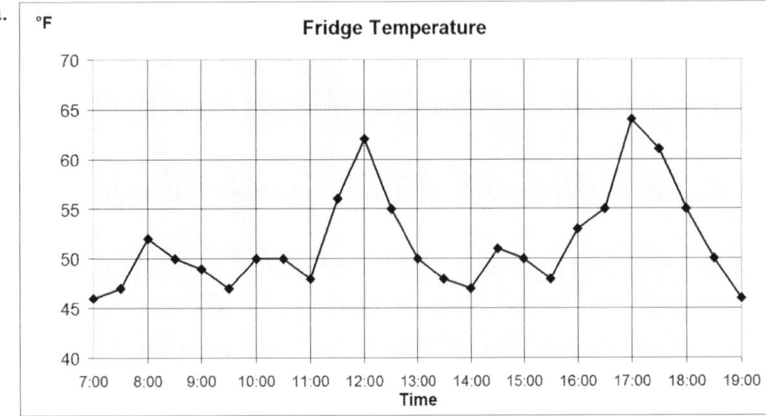

b. The temperature went up. The family probably ate lunch then and had to open the fridge several times before and after eating, which made the fridge temperature go up.

c. The temperature went up again. The family probably ate supper then and had to open the fridge several times before and after eating, which made the fridge temperature go up.

Line Graphs, cont.

3. It is easiest to make the gridlines go by 20s.

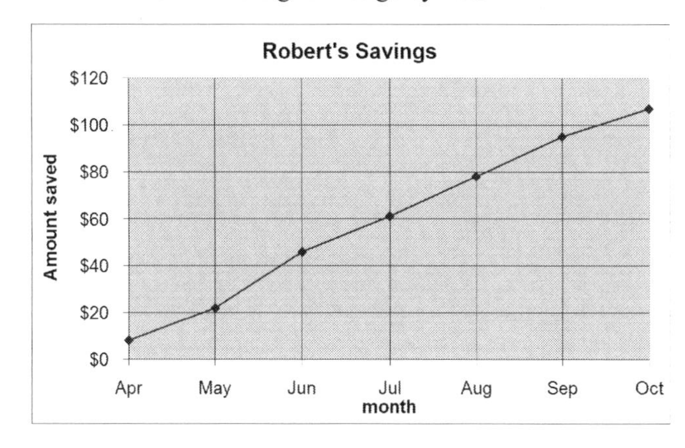

4.

Month	Visitors	rounded to the nearest 50
Jan	1039	1050
Feb	1230	1250
Mar	1442	1450
Apr	1427	1450
May	1183	1200
Jun	823	800
Jul	674	650
Aug	924	900
Sep	1459	1450
Oct	1540	1550
Nov	1638	1650
Dec	1149	1150

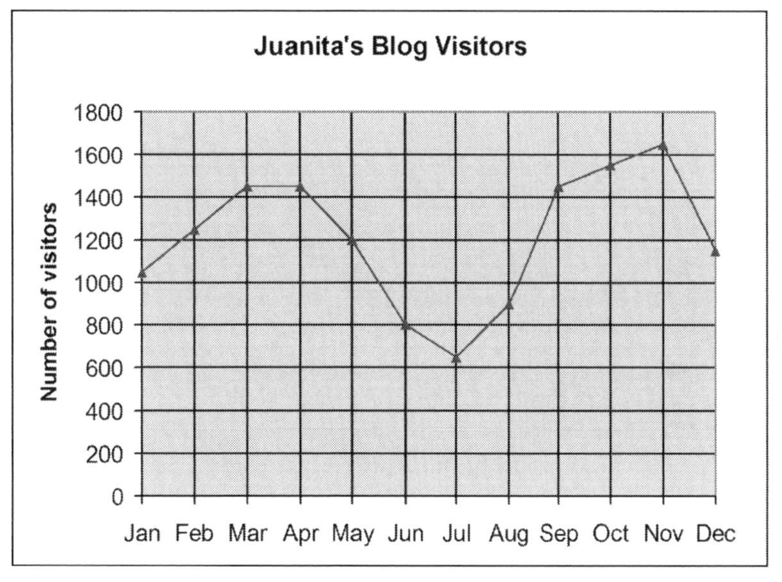

In the _summer_ Juanita's blog had many fewer visitors than in the spring or fall.
The three months with the fewest visitors were _June_, _July_, and _August_.
The three months with the most visitors were _September_, _October_, and _November._
Note: To find the three months with most visitors, look at the actual numbers given in the table.

5. a.

Time	Distance
0 s	0 m
1 s	30 m
2 s	60 m
3 s	90 m
4 s	120 m
5 s	150 m
6 s	180 m
7 s	210 m
8 s	240 m
9 s	270 m
10 s	300 m
11 s	330 m
12 s	360 m

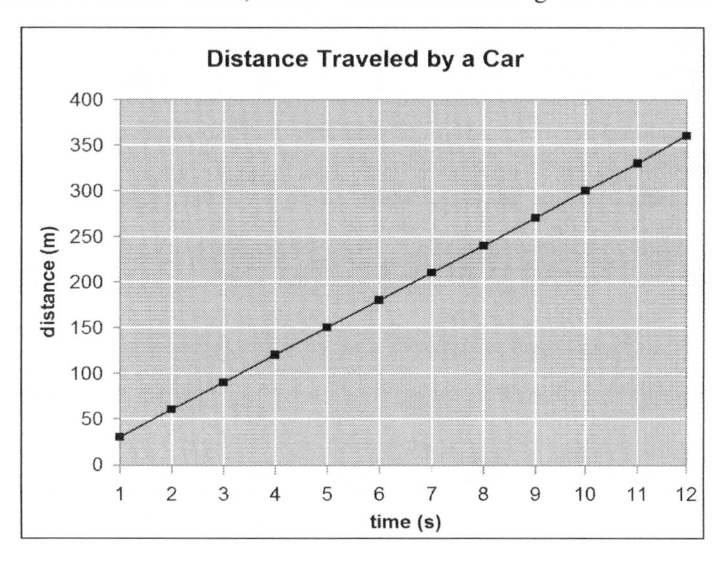

b. The car will have traveled 3 km in 100 seconds, or 1 min 40 s.

Line Graphs, cont.

6. a.

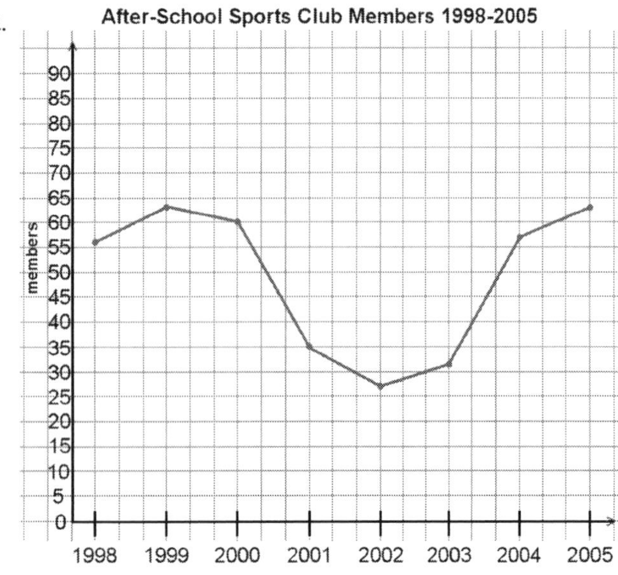

b. Answers may vary. Maybe during 2001-2003 the sports club had a leader that the students didn't like. Or maybe during those years some other activity was offered that was much more popular.

Reading Line Graphs, p. 25

1. a. If you continue the line on the line graph in a similar trend as from 1990 to 2000, the farm population could be about 2,000,000. Of course, this does not guarantee that actually happened. Looking at the numbers given, the decrease from 1990 to 2000 was about 1,500,000 persons. We cannot expect it to drop at the same rate, but perhaps if the farm population dropped at a slower rate, maybe it dropped by a half million or by a million, and was about 1.5 to 2 million people.
   b. In the 1940s and 1950s.
   c. In the 1940s, the farm population decreased by about 7,499,000 people. From 1950 to 1960, it decreased by about 9,603,000 people.
   d. The farm population decreased to under 10 million people in about 1969.
   e. The farm population decreased to under 5 million people in about 1986.

2. a. 293; 807 b. 750; 2,058 c. mammals d. fishes and reptiles e. mammals; from 2002 to 2006

Double and Triple Line Graphs, p. 27

1. a. Mom sent $12 + 6 + 8 + 11 + 5 + 6 + 10 = 58$, and Dad sent $4 + 2 + 2 + 6 + 2 + 3 + 1 = 20$.
   b. Mom sent $11 - 6 = 5$ more messages.
   c. On Sunday the difference was $10 - 1 = 9$ messages.
   d. On Friday $(5 - 2)$ and on Saturday $(6 - 3)$ the difference was only 3 messages.

2. a. In 2005 there were $2 + 4 + 9 + 5 + 7 + 3 + 1 = 31$ storms; in 2006 $1 + 3 + 2 + 4 + 0 = 10$ storms; and in 2007 $1 + 1 + 0 + 3 + 8 + 1 + 0 + 1 = 15$ storms.
   b. The 2005 season was unusually active.
   c. September, since the total number of storms for September is the highest $(5 + 4 + 8 = 17$ storms).

3. Answers will vary.

Double and Triple Line Graphs,cont.

4. a.

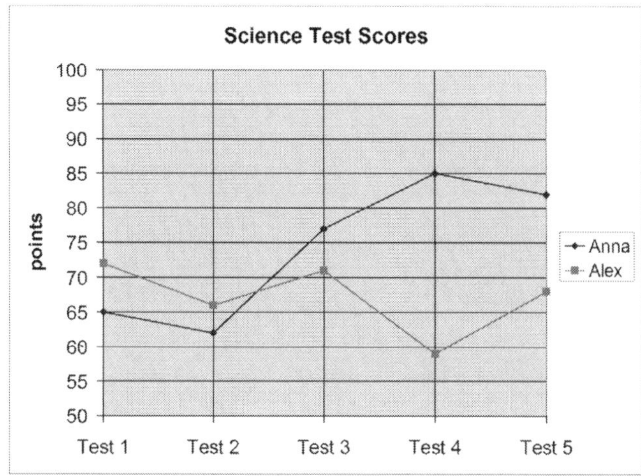

b. Anna improved quite a bit.
c. The difference was the greatest (85 − 59 = 26 points) in test 4.
It was the smallest (66 − 62 = 4 points) in test 2.

Making Bar Graphs, p. 29

1. There are 37,460,000 households that own a cat, and 2,087,000 that own a horse.

2. a. Assuming that the student chooses the scaling for the horizontal axis to go from 0 to 4,000 with tick marks at every 100, the graph will look like the one below. If the student chooses some other scale, such as each tick mark 150 or 200 units apart, then the bars in the graph will appear smaller and there will be lots of empty space in the right part of the graph.

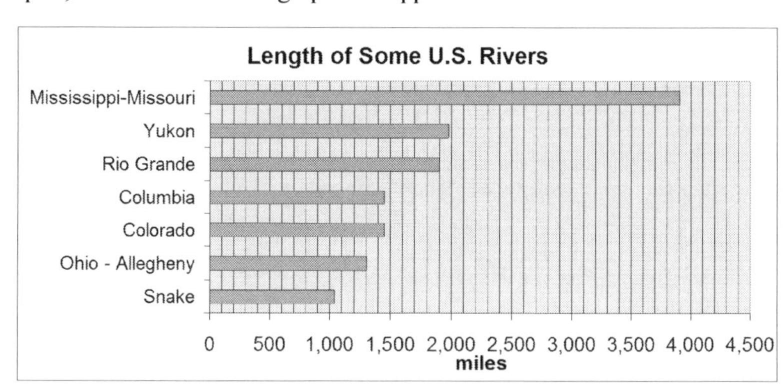

b. About three times as long. c. About two times as long.

3. a.

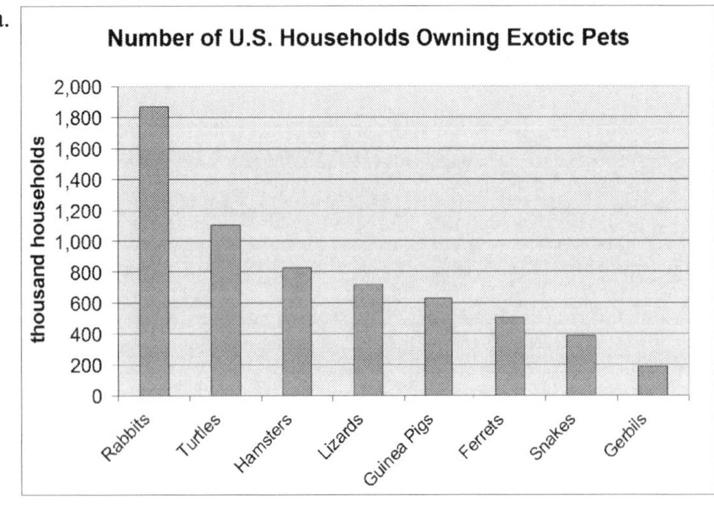

b. About 1,100,000 + 720,000 + 390,000 = 2,210,000 households.

c. The number of households owning a hamster, a guinea pig, or a gerbil is *approximately*: 830,000 + 630,000 + 190,000 = 1,650,000. Since about 2,210,000 households own a turtle, a lizard, or a snake, the latter is more popular.

4.

Number of siblings	frequency
0	3
1	7
2	6
3	2
4	0
5	1
6	1

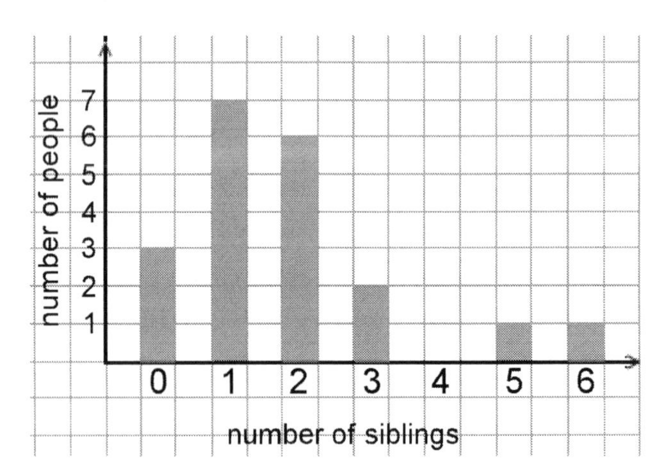

Making Histograms, p. 31

1.

point count	frequency
12-18	2
19-25	5
26-32	6
33-39	3
40-46	3

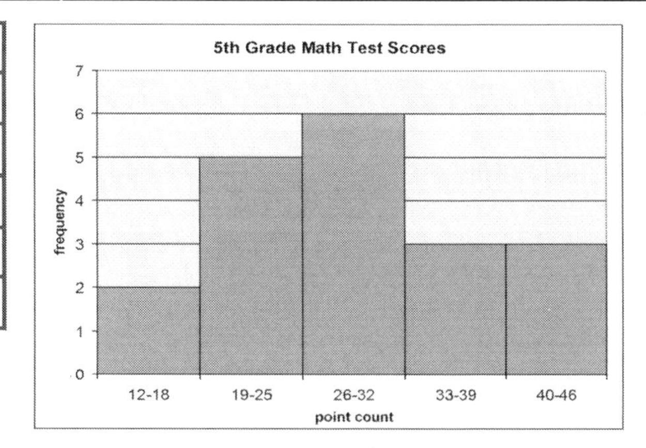

2. The bin width of 4 works well: (73 − 58)/4 = 15/4 = 3.75, rounded up to 4.

weight	frequency
58-61	2
62-65	6
66-69	4
70-73	3

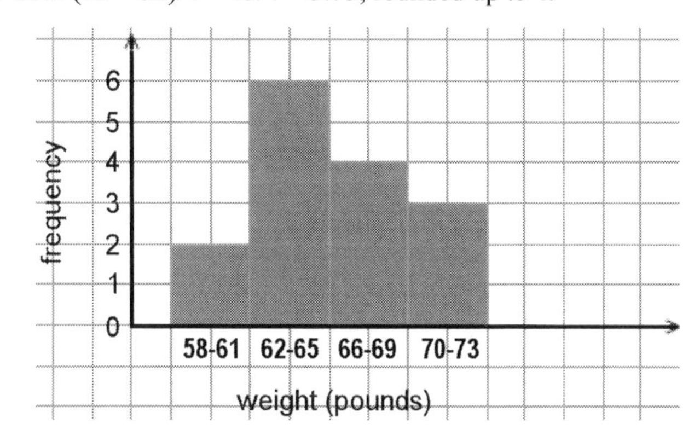

3. The bin width is 9: (43 − 0) / 5 = 8.6, rounded up to 9.

Age	Frequency
0-8	13
9-17	5
18-26	4
27-35	1
36-44	2

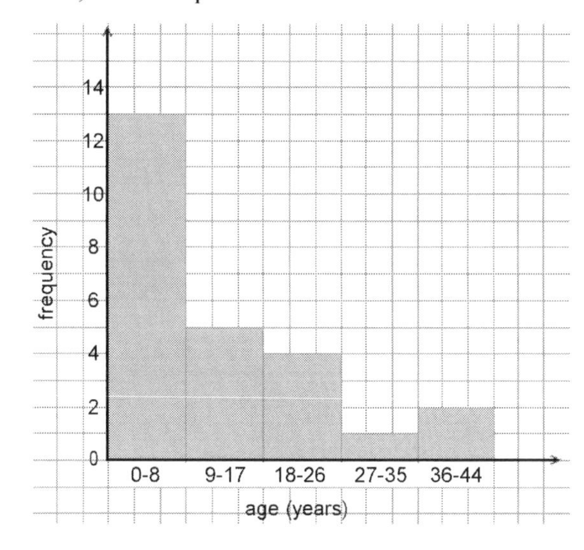

Double Bar graphs, p. 33

1. a.

Grade	can swim	cannot swim
1	20	30
2	26	24
3	38	12
4	46	4
5	48	2

   b. It increases. c. It decreases.

2. a. biographies, mysteries, and poetry

   b. About 25,000 + 32,000 + 26,000 = 83,000 loans.
   (The three most popular genres in 2006 were mysteries, children's, and romance.)

   c. About 13,000 + 7,000 + 6,000 = 26,000 loans
   (The three least popular genres in 2007 were comics, poetry, and biographies.)

3. a.

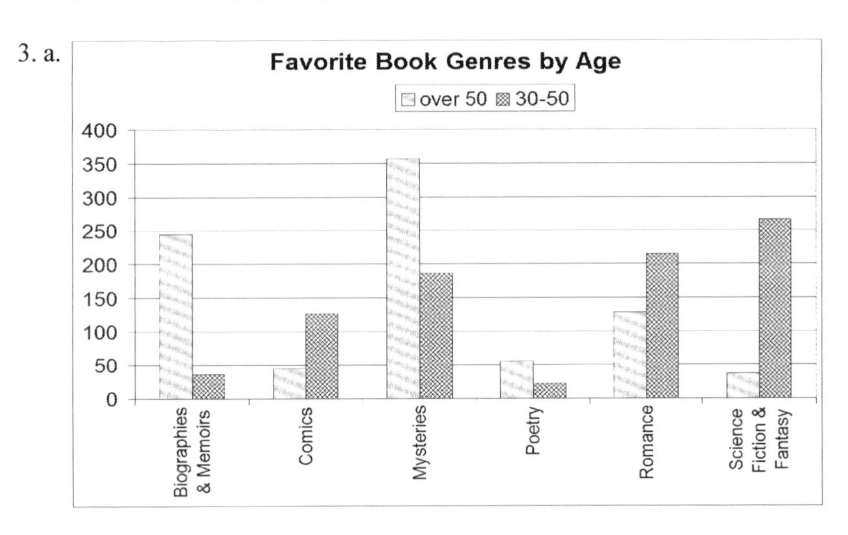

   b. science fiction and fantasy

Average (Mean), p. 35

1. a. 4 b. 16

2. a. 5.4 b. 301.3

3. a. The average is 1,210.7 b. 807 c. 1,446.5

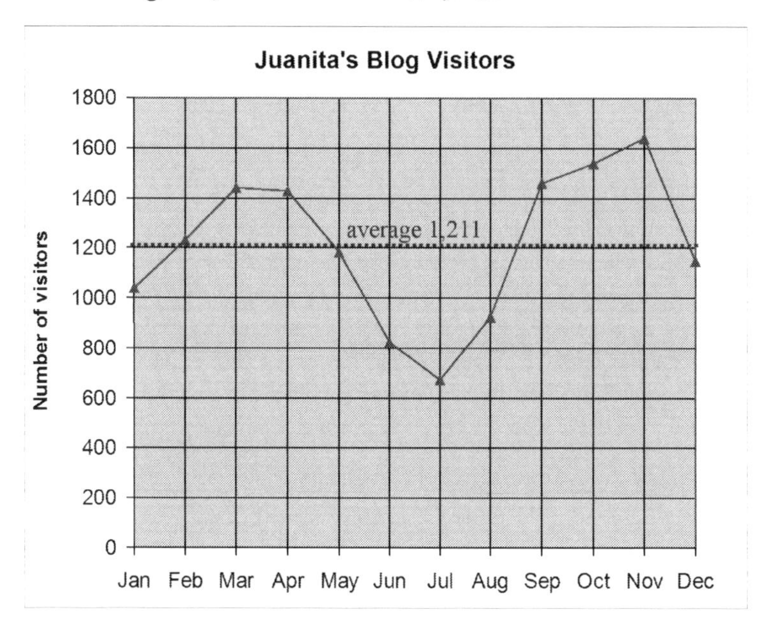

4. The total rainfall calculates to 5.067 mm/day × 15 days = 76.005 mm. Actually, it probably was exactly 76 mm, and the calculated average given in the problem is rounded to three decimals.

5. a. The average weight is 1,527 g.
   b. 1,527 − 1,250 = 277 g c. 1,820 − 1,527 = 293 g
   d. 1,606 g. The average increased by 1,606 − 1,527 = 79 grams

6. a. Mean = $12,969 ÷ 9 = $1,441.
   b. Mean = ($12,969 − $3,400) ÷ 8 = $1,196. The mean decreased by $245 when the highest salary was not included. This shows how "sensitive" the mean can be for small changes in the actual data.

Puzzler corner. $531. Guess and check works here. You can also think logically that since the average is $567, and two of the given prices are higher than the average, then the last unknown price is not more than $567, so it has to be $531.

Mean, Mode, and Bar Graphs, p. 38

1. "Now married"

2. a. "Pop"
   b. "What is your favorite drink?" or "What did you drink yesterday at suppertime?" or "What is your least favorite drink?" *etc.*

3. a. "Vanilla."
   b.

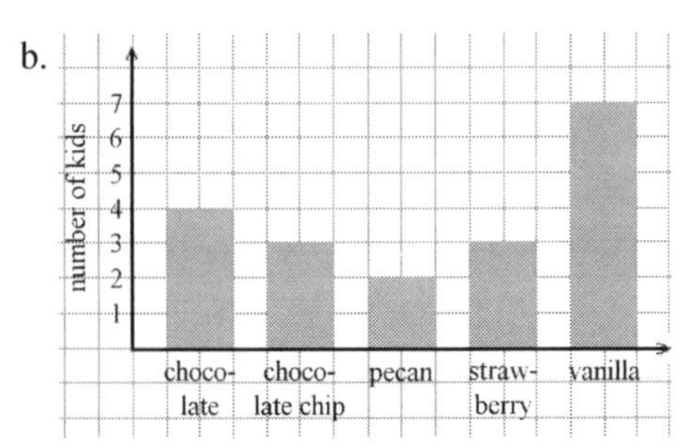

   c. It isn't possible.

4. a. There are three modes: 12, 18, and 19.
   b.

Test Score	Frequency
< 8	3
8..10	4
11..13	6
14..16	1
17..19	7
20..22	1
23..25	2

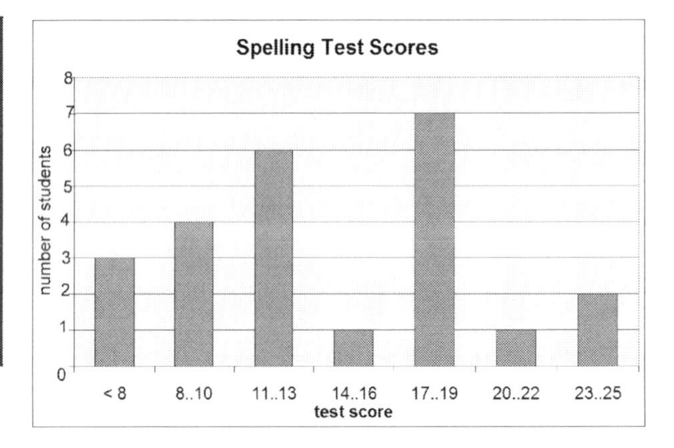

   c. The average is: 336 / 24 = 14

5. a. The mode is "B."
   b.

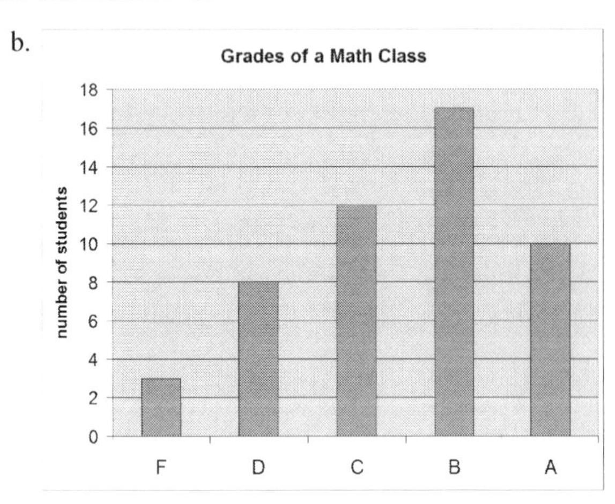

   c. It isn't possible.
   d. There were 50 students in all. 17/50 of the students got a "B."

Statistics Project, p. 40

Answers will vary.

Mixed Review, p. 41

1. a. 1.289 b. 3.108

2.

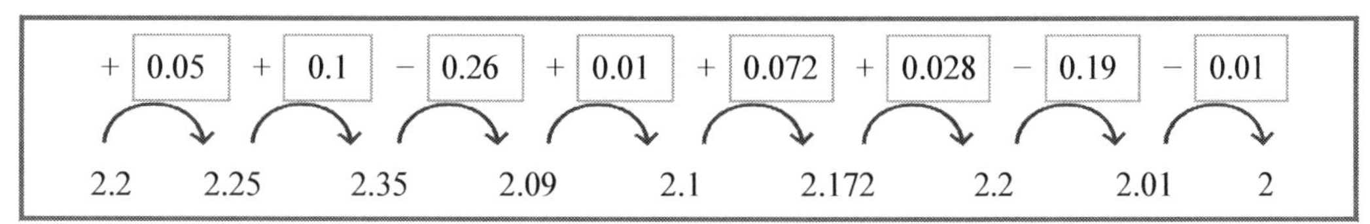

3.

a.	b.	c.	d.
$2 \times 0.06 = 0.12$ $2 \times 0.6 = 1.2$	$0.4 \times 0.7 = 0.28$ $5 \times 0.007 = 0.035$	$100 \times 0.12 = 12$ $0.5 \times 0.03 = 0.015$	$1.1 \times 0.9 = 0.99$ $1000 \times 0.05 = 50$

4. a. Estimate: $4 + 3 + 11 + 2 + 8 \approx \$28$ b. Exact total: \$28.50 c. Error of estimation: \$0.50

5. a. $48 = 2 \times 2 \times 2 \times 2 \times 3$ b. 71 is a prime number c. $93 = 3 \times 31$

6. $x = 5.62$

7. a. 20,000 b. 9,000,000 c. 17,000

8. $x \div 52 = 210$; $x = 10{,}920$

9. $(46 - 6) \div 2 = \$20$; Luisa spent \$20 and Mary spent \$26.

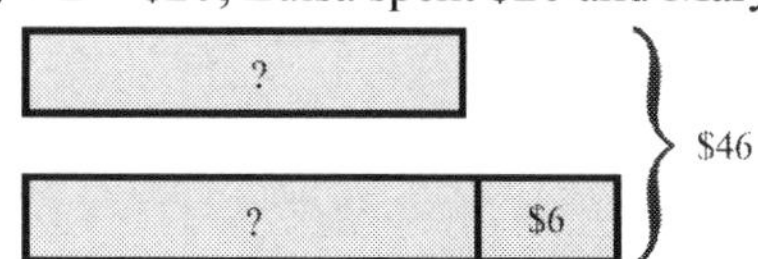

10. Half of John's money is: $2 \times \$48 + \$120 = \$216$. Therefore, he earned $2 \times \$216 = \$432$.

11. Estimations may vary. For example: 12×235 cm $\approx 10 \times 240$ cm $= 2{,}400$ cm $= 24$ m.

12. a. 104 Check: $104 \times 38 = 3{,}952$ b. 15.8 Check: $15.8 \times 17 = 268.6$

13. a. 6.22 b. The division $5.175 \div 0.5$ becomes $51.75 \div 5 = 10.35$.

14.

a. 127,285 + 84,662 (round to thousands)	b. 12,705,143 − 6,460,788 (round to millions)
My estimation: 127,000 + 85,000 = 212,000 Exact answer: 211,947 Error of estimation: 53	My estimation: 13,000,000 − 6,000,000 = 7,000,000 Exact answer: 6,244,355 Error of estimation: 755,645

Review, p. 44

1. The rule is: $y = 9 - x$.

x	0	1	2	3	4
y	9	8	7	6	5

x	5	6	7	8	9
y	4	3	2	1	0

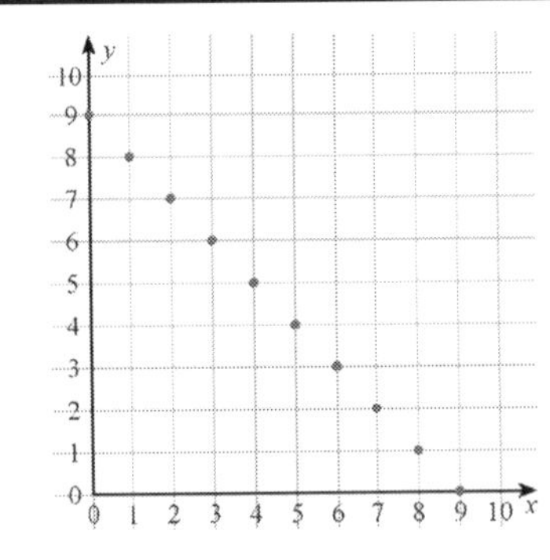

2. The mean is 11.67; the mode is 10.

3. a. Estimate: 3,750,000 tractors in 2010.
   b. From 1940 to 1950, the increase was about 1,750,000 tractors
   c. Slowly declining (but at a slightly increasing rate of decline).
   d. In 1930 there were about 1 million tractors; in 1960 about 4 1/2 million. So the increase was 4 1/2-fold.

4. a. In 2007: June, July, August, and November.
   In 2008: March, May, July, August, and November.
   b. June.

5. a.

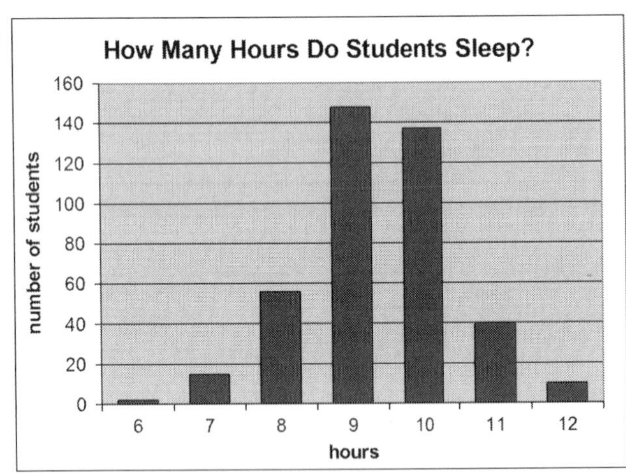

   b. The mode is 9 hours.
   c. The latter (6, 10, 8, 8, 9, 7, 11, 10, 9, 10, 11,...)
   d. 3827 hours ÷ 408 students = 9.38 hours/student ≈ 9.4 hours

Chapter 6: Fractions: Add and Subtract

Review: Mixed Numbers, p. 51

1. a. 1 1/3 b. 2 2/6 c. 3 3/5 d. 6 4/12

2.

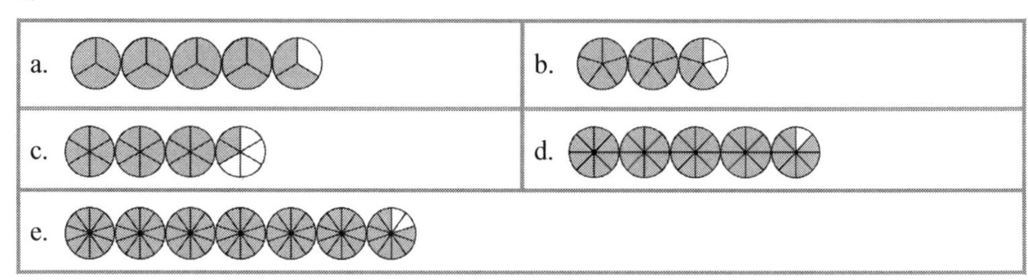

3. a. 2 2/5 b. 1 6/7

4. a. 2/4 b. 1 1/4 c. 3 3/4 d. 4 2/4

5.

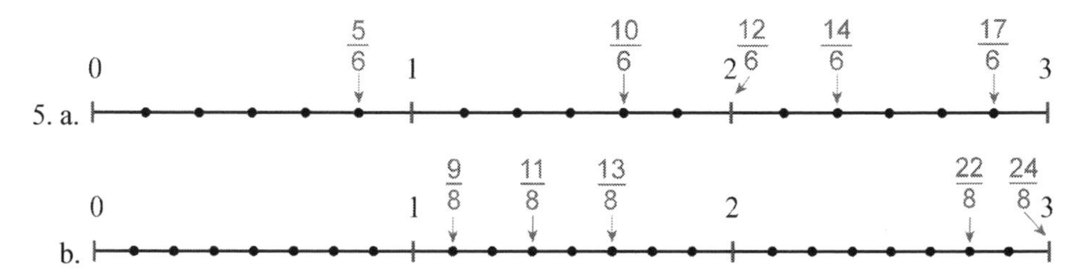

6. a. See the image below. b. 2 2/5 c. See the image below. d. 1 3/5

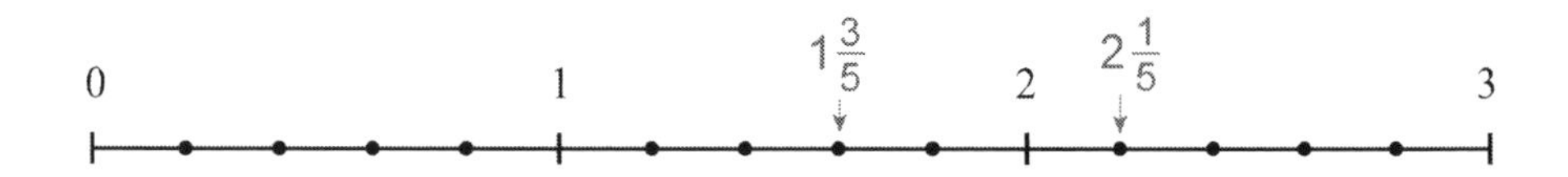

7. a. 1 2/5 = 7/5 b. 2 4/6 = 16/6 c. 2 3/8 = 19/8 d. 4 5/12 = 53/12 e. 3 1/4 = 13/4 f. 5 2/9 = 47/9

8. *There are 5 whole pies, and each pie has 13 slices. So 5 × 13 tells us the number of slices in the whole pies. Then the fractional part 9/13 means that we add 9 slices to that. All total we get 74 slices, and each one is 13th part. So the fraction is* $\frac{74}{13}$.

9. a. 15/2 b. 20/3 c. 75/9 d. 66/10 e. 27/11 f. 97/12 g. 37/16 h. 39/8

10.

a. 47 ÷ 4 = 11 R3 $\frac{47}{4} = 11\frac{3}{4}$	b. 35 ÷ 8 = 4 R 3 $\frac{35}{8} = 4\frac{3}{8}$	c. 19 ÷ 2 = 9 R 1 $\frac{19}{2} = 9\frac{1}{2}$
d. 35 ÷ 6 = 5 R 5 $\frac{35}{6} = 5\frac{5}{6}$	e. 72 ÷ 10 = 7 R 2 $\frac{72}{10} = 7\frac{2}{10}$	f. 22 ÷ 7 = 3 R 1 $\frac{22}{7} = 3\frac{1}{7}$

11. a. 7 6/8 b. 5 1/3 c. 5 2/5 d. 3 5/9 e. 3 1/2 f. 6 1/4 g. 8 2/6 h. 6 2/5 i. 2 2/11 j. 13 k. 7 1/8 l. 9 6/9

Adding Mixed Numbers, p. 55

1\. b. $1\frac{11}{9} \rightarrow 2\frac{2}{9}$ c. $6\frac{7}{4} \rightarrow 7\frac{3}{4}$ d. $3\frac{13}{8} \rightarrow 4\frac{5}{8}$

2\. a. $1\frac{1}{2} + 1\frac{1}{2} + \frac{1}{2} = 3\frac{1}{2}$ b. $2\frac{5}{6} + 1\frac{5}{6} = 4\frac{4}{6}$

c. $2\frac{7}{8} + \frac{3}{8} = 3\frac{2}{8}$ d. $2\frac{1}{7} + \frac{3}{7} + \frac{4}{7} = 3\frac{1}{7}$

e. $2\frac{3}{10} + 1\frac{5}{10} + 1\frac{6}{10} = 5\frac{4}{10}$

3\. a. 12 b. 6 2/5 c. 8 1/9 d. 6 3/7

4\. a. $4\frac{3}{7} + 5\frac{5}{7} = 9\frac{8}{7} \rightarrow 10\frac{1}{7}$ b. $3\frac{3}{5} + 3\frac{4}{5} = 6\frac{7}{5} \rightarrow 7\frac{2}{5}$ c. $4\frac{6}{9} + 2\frac{7}{9} = 6\frac{13}{9} \rightarrow 7\frac{4}{9}$ d. $7\frac{6}{8} + 2\frac{7}{8} = 9\frac{13}{8} \rightarrow 10\frac{5}{8}$

5\. 7 3/8 + 5 7/8 − 1 4/8 = 11 6/8. The combined string is 11 6/8 inches long now.

6\. 1/2 + 1/2 + 3/8 + 3/8 = 1 6/8 cups of flour for a double recipe of recipe 1 and 2 cups of flour for recipe 2.
So, recipe 2 would use 2/8 cup more flour.

7\. a. 1 1/2 b. 2 1/3 c. 3 3/4 d. 5 1/4

8\. a. 5 3/5 b. 3 3/4 c. 9 1/4 d. 5 3/8

9\. a. 6 b. 11 3/5 c. 10 1/8 d. 8 3/10

10\. a. $10\frac{7}{9} + 2\frac{5}{9} + 3\frac{8}{9} = 15\frac{20}{9} \rightarrow 17\frac{2}{9}$ b. $1\frac{5}{11} + 3\frac{9}{11} + 2\frac{8}{11} = 6\frac{22}{11} \rightarrow 8$ c. $2\frac{5}{6} + 5\frac{4}{6} + 2\frac{3}{6} = 9\frac{12}{6} \rightarrow 11$ d. $1\frac{7}{10} + \frac{9}{10} + 10\frac{6}{10} = 11\frac{22}{10} \rightarrow 13\frac{2}{10}$

11\. Jeremy runs 9 miles in a week. Robert runs 10 1/2 miles in a week. Robert runs 1 1/2 miles more in a week.

12\. a. 1 2/4 b. 2 1/5 c. 1 1/3

Subtracting Mixed Numbers 1, p. 58

1. a. 1 7/6 b. 2 9/8 c. 1 11/9 d. 1 8/5 e. 2 13/10 f. 1 5/4

2.

a. $4\frac{2}{9} - 1\frac{8}{9}$ ↓ $= 3\frac{11}{9} - 1\frac{8}{9} = 2\frac{3}{9}$	b. $5\frac{3}{12} - 2\frac{7}{12}$ ↓ $= 4\frac{15}{12} - 2\frac{7}{12} = 2\frac{8}{12}$
c. $5\frac{7}{10} - 3\frac{9}{10}$ ↓ $= 4\frac{17}{10} - 3\frac{9}{10} = 1\frac{8}{10}$	d. $4\frac{3}{8} - 1\frac{7}{8}$ ↓ $= 3\frac{11}{8} - 1\frac{7}{8} = 2\frac{4}{8}$

3.

a. $2\,\frac{13}{9}$ $\cancel{3}\,\cancel{\frac{4}{9}}$ $-\ \frac{8}{9}$ $2\,\frac{5}{9}$	b. $6\,\frac{13}{9}$ $\cancel{7}\,\cancel{\frac{4}{9}}$ $-2\,\frac{7}{9}$ $4\,\frac{6}{9}$	c. $11\,\frac{21}{12}$ $\cancel{12}\,\cancel{\frac{9}{12}}$ $-6\,\frac{11}{12}$ $5\,\frac{10}{12}$	d. $7\,\frac{17}{14}$ $\cancel{8}\,\cancel{\frac{3}{14}}$ $-5\,\frac{9}{14}$ $2\,\frac{8}{14}$
e. $14\,\frac{7}{9}$ $-3\,\frac{5}{9}$ $11\,\frac{2}{9}$	f. $10\,\frac{26}{21}$ $\cancel{11}\,\cancel{\frac{5}{21}}$ $-7\,\frac{15}{21}$ $3\,\frac{11}{21}$	g. $25\,\frac{23}{19}$ $\cancel{26}\,\cancel{\frac{4}{19}}$ $-14\,\frac{15}{19}$ $11\,\frac{8}{19}$	h. $9\,\frac{23}{20}$ $\cancel{10}\,\cancel{\frac{3}{20}}$ $-5\,\frac{7}{20}$ $4\,\frac{16}{20}$

4. a. 1 3/6 b. 3/5 c. 1 2/8 d. 3/7 e. 2 7/9

5. a. $3\frac{1}{4} - \frac{3}{4} = 2\frac{2}{4}$ b. $2\frac{3}{7} - 1\frac{5}{7} = \frac{5}{7}$ c. $3\frac{1}{9} - 1\frac{5}{9} = 1\frac{5}{9}$ d. $2\frac{5}{12} - 1\frac{11}{12} = \frac{6}{12}$

6. a. 4 3/5 b. 2 3/8 c. 2 9/13 d. 4 11/15 e. 3 14/20 f. 3 85/100

7. Subtract 10 1/8 in. − 3 5/8 in. − 3 5/8 in. = 2 7/8 in. The third side of the triangle is 2 7/8 in.

8. 4 yd − 7/8 yd − 7/8 yd = 2 2/8 yd.

9. The two recipes call for 1 3/4 C + 1 3/4 C = 3 2/4 C − 3/4 C = 2 3/4 C. So, he needs 2 3/4 cups more.

Puzzle Corner. The "trick" is to split the pieces further so that they are the same kind of pieces. In this case, sixths will work. So, 2 1/2 becomes 2 3/6.And 1 2/3 becomes 1 4/6.
Now we can subtract: 2 1/2 − 1 2/3 = 2 3/6 − 1 4/6 = 5/6.

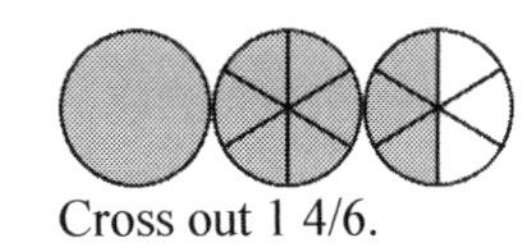

Cross out 1 4/6.

Subtracting Mixed Numbers 2, p. 62

1. a. 3 4/8 b. 3 9/15 c. 3 2/30 d. 1 6/8 e. 9 f. 11 6/12

2. 3 3/4 − 3/4 − 3/4 = 2 1/4; He has 2 1/4 kg of beef left.

3.

a. $5\frac{1}{11} - 3\frac{2}{11} = 1\frac{10}{11}$	b. $6\frac{6}{7} - 1\frac{5}{7} = 5\frac{1}{7}$	c. $6\frac{2}{15} - 1\frac{9}{15} = 4\frac{8}{15}$
$1\frac{10}{11} + 3\frac{2}{11} = 5\frac{1}{11}$	$5\frac{1}{7} + 1\frac{5}{7} = 6\frac{6}{7}$	$4\frac{8}{15} + 1\frac{9}{15} = 6\frac{2}{15}$

4. a. 5 3/5 b. 5 11/12 c. 3 7/9 d. 2 1/3 e. 7 6/12 f. 2 2/6

5.

m. $4\frac{11}{12}$	i. $1\frac{2}{10}$	e. $12\frac{6}{9}$	a. $2\frac{4}{9}$
k. $3\frac{7}{11}$	h. 2		c. $2\frac{4}{11}$
b. $2\frac{12}{15}$			l. $6\frac{6}{8}$
d. $1\frac{7}{9}$	f. $1\frac{3}{11}$	g. $5\frac{6}{12}$	j. $5\frac{4}{8}$

6. a. 4 2/4 b. 9 4/6 c. 8 4/8 d. 10 4/12

Equivalent Fractions 1, p. 64

1.

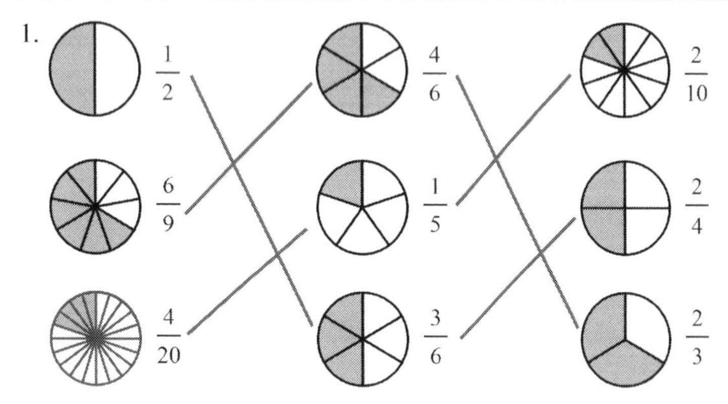

2.

$\frac{1}{2}$	=	$\frac{2}{4}$	=	$\frac{3}{6}$	=	$\frac{4}{8}$	=	$\frac{5}{10}$	=	$\frac{6}{12}$	=	$\frac{7}{14}$	=	$\frac{8}{16}$

Equivalent Fractions 1, cont.

3.

a. Split each piece in two. × 2 $\frac{2}{5} = \frac{4}{10}$ × 2	b. Split each piece into three. × 3 $\frac{1}{2} = \frac{3}{6}$ × 3	c. Split each piece in two. × 2 $\frac{2}{3} = \frac{4}{6}$ × 2
d. Split each piece in two. × 2 $\frac{1}{4} = \frac{2}{8}$ × 2	e. Split each piece into three. × 3 $\frac{3}{3} = \frac{9}{9}$ × 3	f. Split each piece in two. × 2 $\frac{1}{5} = \frac{2}{10}$ × 2
g. Split each piece in two. × 2 $\frac{1}{2} = \frac{2}{4}$ × 2	h. Split each piece in two. × 2 $\frac{3}{8} = \frac{6}{16}$ × 2	i. Split each piece into five. × 5 $\frac{1}{2} = \frac{5}{10}$ × 5

4.

a.	b.	c.	d.	e.
$\frac{3}{4} = \frac{12}{16}$	$\frac{5}{8} = \frac{10}{16}$	$\frac{1}{2} = \frac{6}{12}$	$\frac{2}{7} = \frac{8}{28}$	$\frac{1}{4} = \frac{5}{20}$
f.	g.	h.	i.	j.
$\frac{2}{7} = \frac{6}{21}$	$\frac{5}{8} = \frac{50}{80}$	$\frac{1}{2} = \frac{8}{16}$	$\frac{3}{5} = \frac{21}{35}$	$\frac{3}{7} = \frac{24}{56}$

5.

a. Pieces were split into three. × 3 $\frac{4}{7} = \frac{12}{21}$ × 3	b. Pieces were split into four. × 4 $\frac{4}{5} = \frac{16}{20}$ × 4	c. Pieces were split into three. × 3 $\frac{1}{6} = \frac{3}{18}$ × 3	d. Pieces were split into two. × 2 $\frac{6}{7} = \frac{12}{14}$ × 2	e. Pieces were split into four. × 4 $\frac{2}{3} = \frac{8}{12}$ × 4
f. $\frac{7}{10} = \frac{14}{20}$	g. $\frac{5}{9} = \frac{15}{27}$	h. $\frac{1}{8} = \frac{6}{48}$	i. $\frac{4}{9} = \frac{24}{54}$	j. $\frac{8}{11} = \frac{32}{44}$
k. $\frac{3}{10} = \frac{9}{30}$	l. $\frac{2}{11} = \frac{6}{33}$	m. $\frac{4}{7} = \frac{32}{56}$	n. $\frac{1}{6} = \frac{9}{54}$	o. $\frac{7}{8} = \frac{56}{64}$

Equivalent Fractions 1, cont.

6. a. 2/3 = 8/12 = 16/24
   b. 5/6 = 10/12 = 20/24
   c. Answers vary; any of the following will do:
   1/12, 2/12, 3/12, 5/12, 6/12, 7/12, 9/12, 10/12, or 11/12.
   d. Answers vary; any of the following will do:
   1/24, 3/24, 5/24, 7/24, 9/24, 11/24, 13/24, 15/24,
   17/24, 19/24, 21/24, or 23/24.

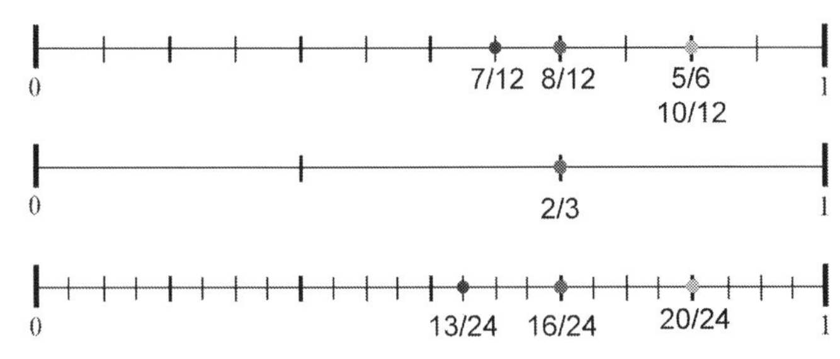

7. a. & b.

	Fraction	Equivalent fraction
Dad	1/2	6/12
Mom	1/3	4/12
Cindy	1/3	4/12
Derek		10/12

b. Derek ate 10/12 of a pizza. Note: the total comes to 24/12, which is equal to two pizzas.

Equivalent Fractions 2, p. 67

1.

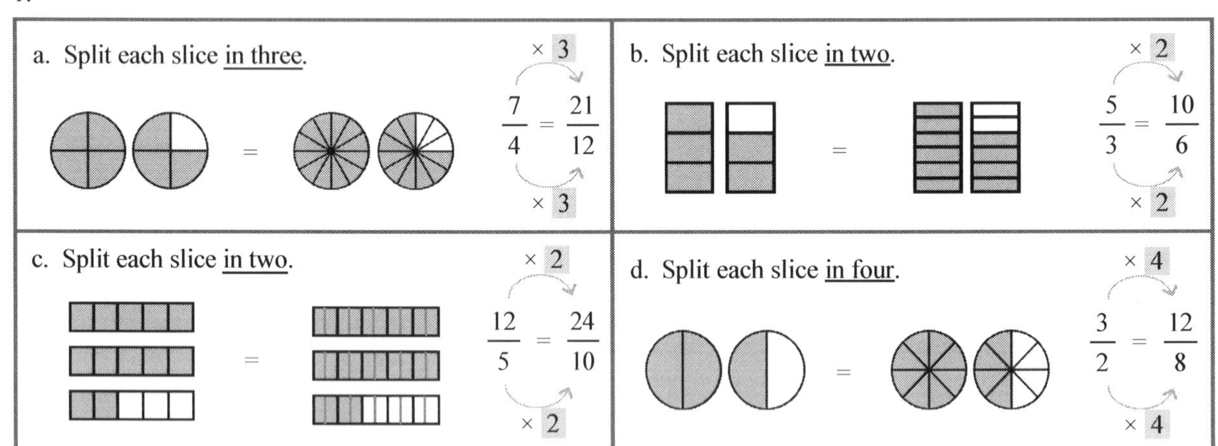

2. a. 1 12/16 b. 5 28/40 c. 36/16 d. 3 12/18 e. 40/15 f. 21/6 g. 6 12/54 h. 42/6 i. 5 56/80 j. 54/18

3.

whole pies	halves	thirds	fourths	fifths	tenths	hundredths
$\frac{3}{1}$	$\frac{6}{2}$	$\frac{9}{3}$	$\frac{12}{4}$	$\frac{15}{5}$	$\frac{30}{10}$	$\frac{300}{100}$

4.

halves	fourths	sixths	eighths	tenths	twentieths	hundredths
$\frac{5}{2}$	$\frac{10}{4}$	$\frac{15}{6}$	$\frac{20}{8}$	$\frac{25}{10}$	$\frac{50}{20}$	$\frac{250}{100}$

Equivalent Fractions 2, cont.

5.

a. $\frac{5}{7} = \frac{20}{28}$ The pieces were split into _4_.	b. NOT POSSIBLE	c. NOT POSSIBLE	d. $\frac{2}{3} = \frac{8}{12}$ The pieces were split into _4_.	e. NOT POSSIBLE
f. NOT POSSIBLE	g. $\frac{2}{9} = \frac{14}{63}$ The pieces were split into _7_.	h. $\frac{5}{4} = \frac{40}{32}$ The pieces were split into _8_.	i. $\frac{1}{3} = \frac{5}{15}$ The pieces were split into _5_.	j. NOT POSSIBLE

6. Answers vary. If we know the new numerator is not divisible by the old one, or if the new denominator is not divisible by the old one, then the conversion is not possible. In other words, if the numerator does not "go into" or divide into the new numerator and similarly with the denominators, then we cannot find an equivalent fraction.

7.

a. $\frac{3}{4} = \frac{6}{8} = \frac{9}{12} = \frac{12}{16} = \frac{15}{20} = \frac{18}{24} = \frac{21}{28} = \frac{24}{32} = \frac{27}{36}$
b. $\frac{5}{3} = \frac{10}{6} = \frac{15}{9} = \frac{20}{12} = \frac{25}{15} = \frac{30}{18} = \frac{35}{21} = \frac{40}{24} = \frac{45}{27}$

Adding and Subtracting Unlike Fractions, p. 69

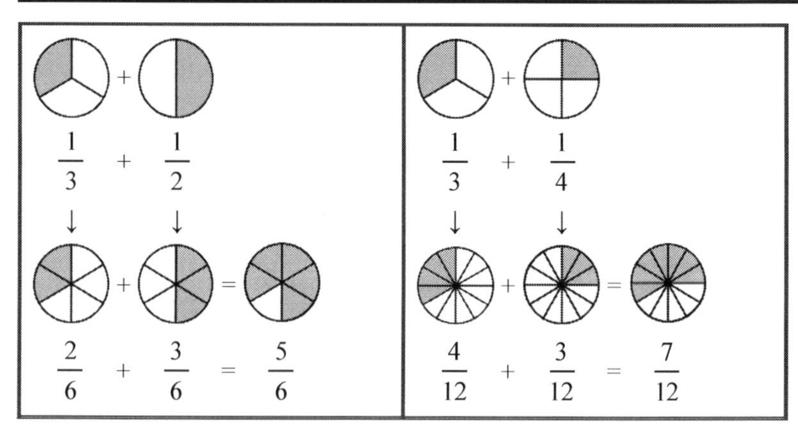

1.

a. $\frac{1}{2} + \frac{1}{4}$ ↓ ↓ $\frac{2}{4} + \frac{1}{4} = \frac{3}{4}$	b. $\frac{2}{5} + \frac{1}{2}$ ↓ ↓ $\frac{4}{10} + \frac{5}{10} = \frac{9}{10}$	c. $\frac{3}{9} + \frac{1}{3}$ ↓ ↓ $\frac{3}{9} + \frac{3}{9} = \frac{6}{9}$

Adding and Subtracting Unlike Fractions, cont.

2.

a. $\frac{1}{2} + \frac{1}{6}$ $\downarrow \quad \downarrow$ $\frac{3}{6} + \frac{1}{6} = \frac{4}{6}$	b. $\frac{1}{8} + \frac{1}{4}$ $\downarrow \quad \downarrow$ $\frac{1}{8} + \frac{2}{8} = \frac{3}{8}$	c. $\frac{1}{6} + \frac{1}{4}$ $\downarrow \quad \downarrow$ $\frac{2}{12} + \frac{3}{12} = \frac{5}{12}$
d. $\frac{5}{6} - \frac{1}{2}$ $\downarrow \quad \downarrow$ $\frac{5}{6} - \frac{3}{6} = \frac{2}{6}$	e. $\frac{5}{8} - \frac{1}{4}$ $\downarrow \quad \downarrow$ $\frac{5}{8} - \frac{2}{8} = \frac{3}{8}$	f. $\frac{5}{6} - \frac{1}{4}$ $\downarrow \quad \downarrow$ $\frac{10}{12} - \frac{3}{12} = \frac{7}{12}$
g. $\frac{1}{2} + \frac{1}{8}$ $\downarrow \quad \downarrow$ $\frac{4}{8} + \frac{1}{8} = \frac{5}{8}$	h. $\frac{3}{10} + \frac{1}{5}$ $\downarrow \quad \downarrow$ $\frac{3}{10} + \frac{2}{10} = \frac{5}{10}$	i. $\frac{2}{5} + \frac{1}{2}$ $\downarrow \quad \downarrow$ $\frac{4}{10} + \frac{5}{10} = \frac{9}{10}$
j. $\frac{1}{2} + \frac{3}{8}$ $\downarrow \quad \downarrow$ $\frac{4}{8} + \frac{3}{8} = \frac{7}{8}$	k. $\frac{9}{10} - \frac{2}{5}$ $\downarrow \quad \downarrow$ $\frac{9}{10} - \frac{4}{10} = \frac{5}{10}$	l. $\frac{4}{5} - \frac{1}{2}$ $\downarrow \quad \downarrow$ $\frac{8}{10} - \frac{5}{10} = \frac{3}{10}$

3.

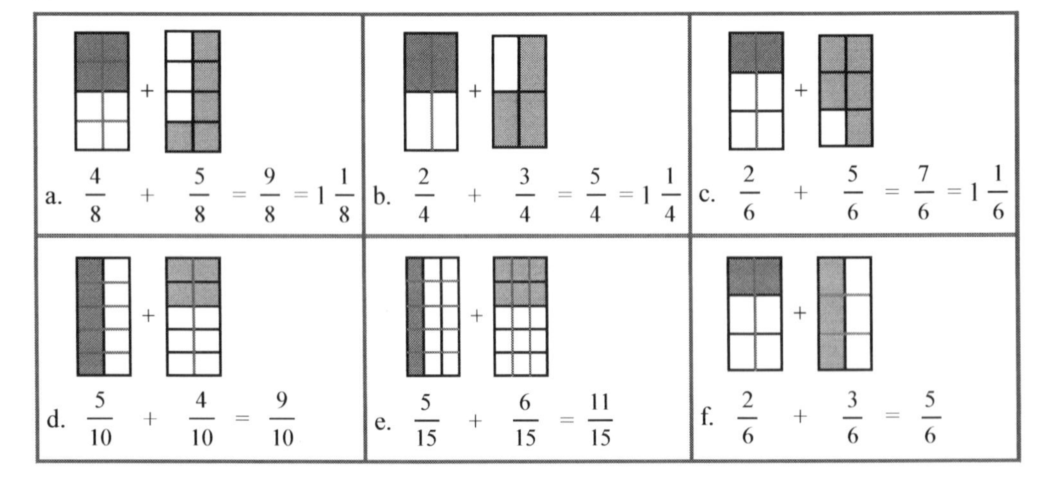

4. a. 8th parts b. 4th parts c. 6th parts d. 10th parts e. 15th parts f. 6th parts

5. The two denominators always "go into" the number that tells us what kind of parts we are converting to. In other words, we need to find a number that is divisible by the two denominators, or in yet other words, a number that is a multiple of both of the denominators.

6.

a. $\frac{1}{2} + \frac{2}{3}$ $\downarrow \quad \downarrow$ $\frac{3}{6} + \frac{4}{6} = \frac{7}{6} = 1\frac{1}{6}$	b. $\frac{2}{3} - \frac{2}{5}$ $\downarrow \quad \downarrow$ $\frac{10}{15} - \frac{6}{15} = \frac{4}{15}$	c. $\frac{1}{3} + \frac{3}{4}$ $\downarrow \quad \downarrow$ $\frac{4}{12} + \frac{9}{12} = \frac{13}{12} = 1\frac{1}{12}$

Finding the (Least) Common Denominator, p. 72

1.

a. $\frac{1}{3} + \frac{3}{5}$ ↓ ↓ $\frac{5}{15} + \frac{9}{15} = \frac{14}{15}$	b. $\frac{6}{7} - \frac{1}{2}$ ↓ ↓ $\frac{12}{14} - \frac{7}{14} = \frac{5}{14}$	c. $\frac{1}{6} + \frac{2}{5}$ ↓ ↓ $\frac{5}{30} + \frac{12}{30} = \frac{17}{30}$
d. $\frac{5}{9} - \frac{1}{3}$ ↓ ↓ $\frac{5}{9} - \frac{3}{9} = \frac{2}{9}$	e. $\frac{1}{8} + \frac{3}{4}$ ↓ ↓ $\frac{1}{8} + \frac{6}{8} = \frac{7}{8}$	f. $\frac{5}{7} - \frac{2}{3}$ ↓ ↓ $\frac{15}{21} - \frac{14}{21} = \frac{1}{21}$
g. $\frac{2}{5} + \frac{1}{4}$ ↓ ↓ $\frac{8}{20} + \frac{5}{20} = \frac{13}{20}$	h. $\frac{5}{6} - \frac{3}{4}$ ↓ ↓ $\frac{10}{12} - \frac{9}{12} = \frac{1}{12}$	i. $\frac{3}{4} - \frac{3}{7}$ ↓ ↓ $\frac{21}{28} - \frac{12}{28} = \frac{9}{28}$

2. a. 20 b. 21 c. 10 (20 is okay, too, though not the best)
d. 12 (24 and 48 are okay too, though not the best) e. 14 f. 18 (36 and 54 are okay too, though not the best)

3.

a. $\frac{4}{5} + \frac{1}{4}$ ↓ ↓ $\frac{16}{20} + \frac{5}{20} = \frac{21}{20}$	b. $\frac{2}{3} - \frac{1}{7}$ ↓ ↓ $\frac{14}{21} - \frac{3}{21} = \frac{11}{21}$	c. $\frac{3}{10} + \frac{1}{2}$ ↓ ↓ $\frac{3}{10} + \frac{5}{10} = \frac{8}{10}$
d. $\frac{4}{12} + \frac{1}{4}$ ↓ ↓ $\frac{4}{12} + \frac{3}{12} = \frac{7}{12}$	e. $\frac{1}{2} - \frac{2}{7}$ ↓ ↓ $\frac{7}{14} - \frac{4}{14} = \frac{3}{14}$	f. $\frac{5}{6} - \frac{4}{9}$ ↓ ↓ $\frac{15}{18} - \frac{8}{18} = \frac{7}{18}$

4. a. LCD is 24. 10/24 + 9/24 = 19/24 b. LCD is 44. 77/44 − 36/44 = 41/44
c. LCD is 36. 3/36 + 4/36 = 7/36 d. LCD is 72. 63/72 − 32/72 = 31/72

Add and Subtract: More Practice, p. 75

1.

a. Amanda added: $\frac{5}{8} + \frac{5}{8} = \frac{10}{16}$

How can you tell it is wrong?

Answers may vary. For example:
The answer, 10/16, is equal to 5/8!
(They are equivalent fractions.)

Or, because we are adding like fractions, which is easy: you simply add the number of pieces. We have here 5 and 5 pieces, or 10 pieces, so the answer is 10 eighths.

Correct answer: 10/8 = 1 2/8 or 1 1/4

b. Robert subtracted: $\frac{7}{9} - \frac{1}{2} = \frac{6}{7}$

How can you tell it is wrong?

Answers may vary. For example:
7/9 is nearly 1, and we subtract half from it. The answer, 6/7, is also nearly 1, so it does not make sense.

Correct answer: 14/18 − 9/18 = 5/18

2.

Olivia subtracted: $\frac{7}{12} - \frac{1}{4} = \frac{1}{2}$

Correct answer: 7/12 − 3/12 = 4/12 (which is 1/3.)

How can you tell it is wrong?

Answers may vary. For example: 7/12 is close to one-half. We subtract 1/4 from it —and the answer says we get 1/2! That does not make sense.

3.

$\frac{23}{36}$	$\frac{5}{6}$	$\frac{7}{15}$	$\frac{3}{10}$	$\frac{1}{10}$	$\frac{4}{15}$	$\frac{5}{6}$		$\frac{23}{30}$	$\frac{7}{45}$		$\frac{8}{15}$	$\frac{31}{30}$	$\frac{17}{28}$
B	E	C	A	U	S	E		I	T		W	A	S

$\frac{17}{21}$	$\frac{9}{10}$	$\frac{9}{35}$		$\frac{83}{72}$	$\frac{11}{24}$	$\frac{11}{24}$	$\frac{27}{40}$	$\frac{23}{30}$	$\frac{5}{9}$	$\frac{1}{6}$		$\frac{1}{14}$	$\frac{104}{63}$	$\frac{7}{6}$	$\frac{13}{20}$	
N	O	T		P	E	E	L	I	N	G		W	E	L	L	!

Puzzle Corner

$\frac{2}{3}$	+	$\frac{1}{5}$	=	$\frac{13}{15}$
+		+		
$\frac{1}{6}$	+	$\frac{1}{4}$	=	$\frac{5}{12}$
=		=		
$\frac{5}{6}$		$\frac{9}{20}$		

$\frac{1}{6}$	+	$\frac{1}{7}$	=	$\frac{13}{42}$
+		+		
$\frac{1}{8}$	+	$\frac{1}{9}$	=	$\frac{17}{72}$
=		=		
$\frac{7}{24}$		$\frac{16}{63}$		

Adding and Subtracting Mixed Numbers, p. 78

1.

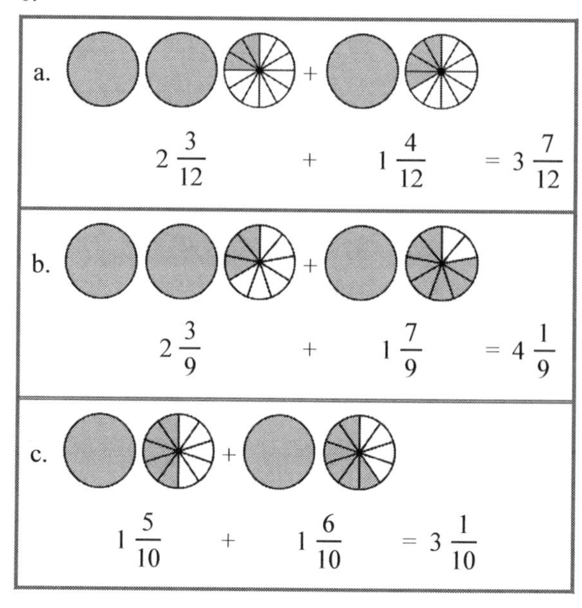

2.

a. $6\frac{2}{3} + 3\frac{1}{5} \Rightarrow 6\frac{10}{15} + 3\frac{3}{15} = 9\frac{13}{15}$

b. $10\frac{1}{8} + 3\frac{2}{5} \Rightarrow 10\frac{5}{40} + 3\frac{16}{40} = 13\frac{21}{40}$

c. $17\frac{1}{16} + 3\frac{3}{8} \Rightarrow 17\frac{1}{16} + 3\frac{6}{16} = 20\frac{7}{16}$

3.

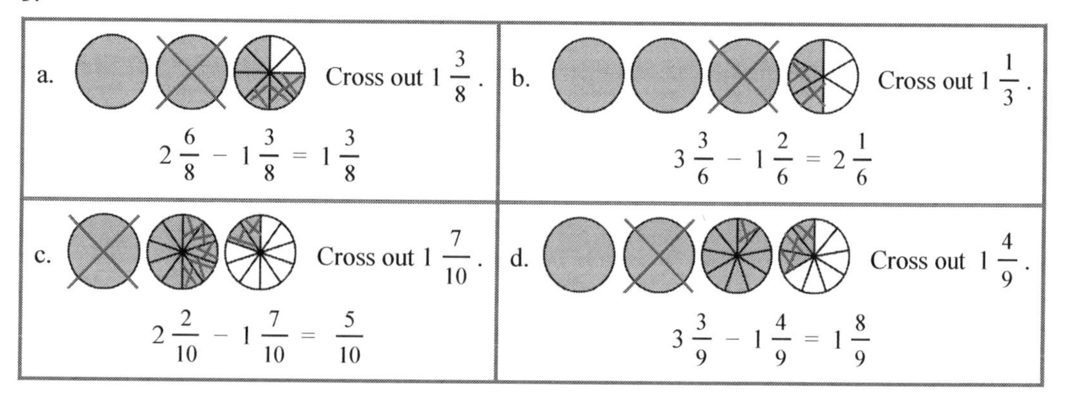

4.

a. $5\frac{1}{2} \Rightarrow 4\frac{15}{10}$ $-\ 2\frac{4}{5} \qquad -\ 2\frac{8}{10}$ $2\frac{7}{10}$	b. $15\frac{4}{8} \Rightarrow 14\frac{36}{24}$ $-\ 8\frac{5}{6} \qquad -\ 8\frac{20}{24}$ $6\frac{16}{24}$	c. $16\frac{5}{9} \Rightarrow 16\frac{10}{18}$ $-\ 10\frac{1}{2} \qquad -\ 10\frac{9}{18}$ $6\frac{1}{18}$
d. $4\frac{1}{6} \Rightarrow 3\frac{35}{30}$ $-\ 2\frac{3}{5} \qquad -\ 2\frac{18}{30}$ $1\frac{17}{30}$	e. $11\frac{1}{12} \Rightarrow 10\frac{13}{12}$ $-\ 3\frac{1}{4} \qquad -\ 3\frac{3}{12}$ $7\frac{10}{12}$	f. $8\frac{2}{9} \Rightarrow 7\frac{44}{36}$ $-\ 2\frac{3}{4} \qquad -\ 2\frac{27}{36}$ $5\frac{17}{36}$

5.

a. $4\frac{1}{2} \Rightarrow 4\frac{5}{10}$ $+\ 3\frac{4}{5} \qquad +\ 3\frac{8}{10}$ $7\frac{13}{10} \Rightarrow 8\frac{3}{10}$	b. $5\frac{5}{6} \Rightarrow 5\frac{5}{6}$ $+\ 7\frac{2}{3} \qquad +\ 7\frac{4}{6}$ $12\frac{9}{6} \Rightarrow 13\frac{3}{6}$
c. $3\frac{5}{6} \Rightarrow 3\frac{20}{24}$ $+\ 2\frac{7}{8} \qquad +\ 2\frac{21}{24}$ $5\frac{41}{24} \Rightarrow 6\frac{17}{24}$	d. $9\frac{5}{7} \Rightarrow 9\frac{15}{21}$ $+\ 7\frac{2}{3} \qquad +\ 7\frac{14}{21}$ $16\frac{29}{21} \Rightarrow 17\frac{8}{21}$

6. a. 5 6/8 − 1 7/8 = 3 7/8 b. 8 9/15 + 5 12/15 = 14 6/15 c. 3 2/9 − 1 3/9 = 1 8/9
d. 7 4/14 − 2 7/14 = 4 11/14 e. 8 3/10 + 2 8/10 = 11 1/10 f. 6 14/21 − 1 3/21 = 5 11/21

7. a. 1 3/4 lb + 1 2/3 lb = 1 9/12 lb + 1 8/12 lb = 3 5/12 lb
b. 1 3/4 lb = 1 lb 12 oz; 1 2/3 lb = 1 lb 10.7 oz
c. 3 lb 6.7 oz
d. Answers will vary.

8. a. 1 1/4 m + 8/10 m = 1 5/20 m + 16/20 m = 2 1/20 m
b. 1 1/4 m = 1 m 25 cm = 125 cm; 8/10 m = 80 cm. The total length is 205 cm.

9. 1 − 1/10 − 1/2 = 1 − 1/10 − 5/10 = 4/10. Jerry would do 4/10 of the project.

10. 3 1/2 dl + 5 dl + 3/4 dl = 9 1/4 dl. Yes, 1 kg of flour is enough.

11. a. 1 − 19/100 − 2/10 = 1 − 19/100 − 20/100 = 61/100. So, 61/100 of his salary is left.
b. 1/100 of $2,000 is $20. So, 61/100 of it is 61 × $20 = $1,220.

12. a. 1/8 + 1/8 + 1/4 + 1/4 = 1/8 + 1/8 + 2/8 + 2/8 = 6/8. Or, you can note that 1/8 + 1/8 makes 1/4, so in total they drank 3/4 of the smoothie. So, 2/8 or 1/4 of it is left.
b. 1/2 liter

13. The margins are a total of 3/4 in. + 3/4 in. = 1 1/2 inches. Subtract that from the width and height of the notebook. The width of the picture will be 3 1/4 in. − 1 1/2 in. = 1 3/4 in. and the height will be 6 1/8 in. − 1 1/2 in. = 4 5/8 in.

Comparing Fractions, p. 83

1. a. > b. < c. > d. < e. > f. > g. < h. > i. < j. <
k. = l. < m. < n. > o. < p. < q. > r. > s. > t. =

2. a. 16/24 > 15/24 b. 20/24 < 21/24 c. 10/30 > 9/30 d. 40/60 < 50/70
e. 15/24 > 14/24 f. 55/40 < 56/40 g. 60/100 > 58/100 h. 54/45 < 55/45
i. 49/70 < 50/50 j. 43/100 > 30/100 k. 63/56 < 64/56 l. 21/30 > 20/30

3. 1 1/2 cups is more than 1 1/3 cups so a triple batch of the first recipe uses more sugar - 1/6 cup more

4. a. > b. > c. > d. < e. < f. > g. < h. < i. > j. > k. < l. >

5. Since 1/4 < 3/10, taking off 3/10 of the price is bigger.
The answer does not change if the price changes: 3/10 off of any price is a greater discount than 1/4 off of the same price.

6. a. $\frac{1}{5} < \frac{1}{3} < \frac{2}{5} < \frac{1}{2} < \frac{2}{3}$ b. $\frac{2}{2} < \frac{6}{5} < \frac{4}{3} < \frac{7}{5} < \frac{3}{2}$

7. a. $\frac{7}{9} < \frac{7}{8} < \frac{9}{10}$ b. $\frac{2}{9} < \frac{1}{3} < \frac{4}{10}$

8. a. The women who swam were the bigger group. One-hundredth of 600 is 6; therefore 22/100 of 600 is $22 \times 6 = 132$.
One-fifth of 600 is 120.

b. Since 1/3 of them never exercise, 2/3 exercise.

c. Four hundred women of that group exercise.

d. There are 132 women who swim. (See the answer for (a).)

Puzzle corner. Either way, the dog would get 3/24 (or 1/8) of the pizza.

Measuring in Inches, p. 87

1\.

1/2 1 1/2 2 1/2

0 1 2 3

1/2 1 1/2 2 1/2

0 1 2 3

1/2 1 1/2 2 1/2

0 1 2 3

1/2 1 1/2 2 1/2

0 1 2 3

2\.

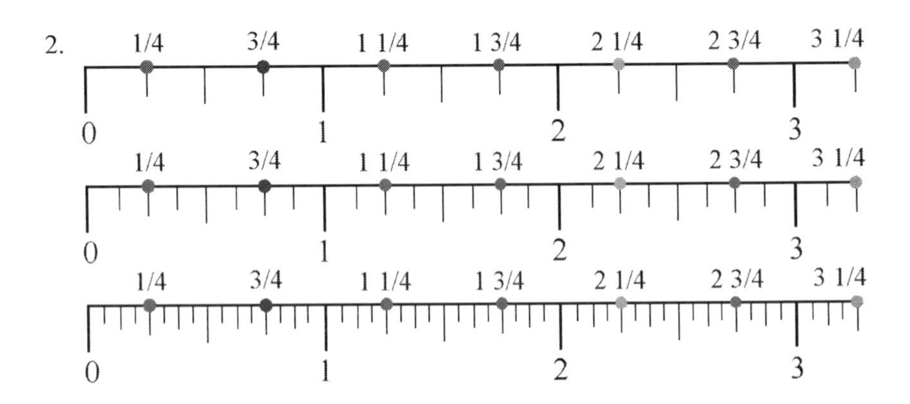

3\.

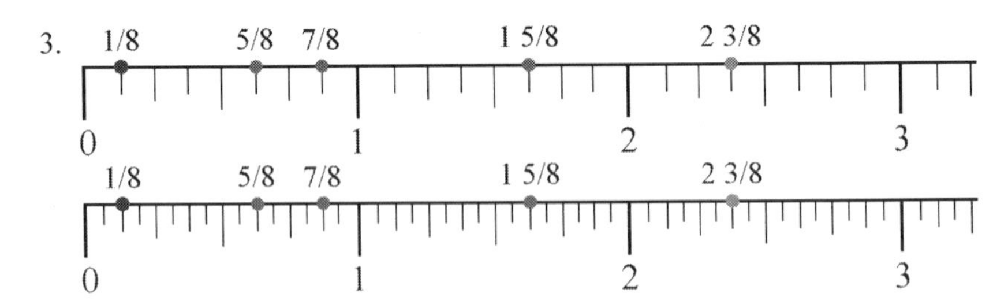

4\.

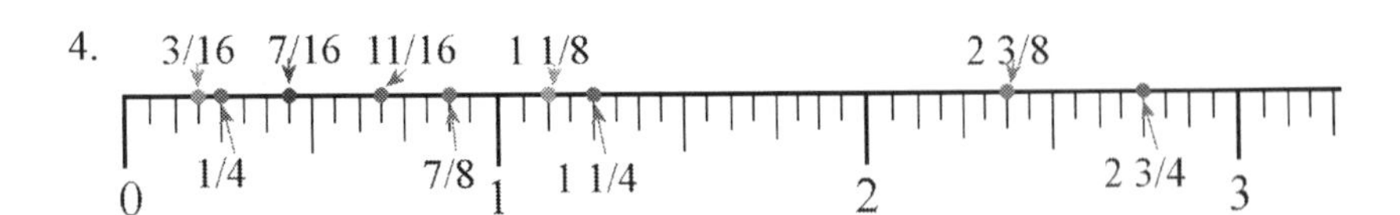

5\. a. 1 1/4 in.
b. 3/4 in.
c. 1 2/8 in. or 1 3/8 in.
d. 5/8 in. or 6/8 in.
e. 1 5/16 in.
f. 11/16 in.
g. 2 3/4 in.
h. 2 5/8 in.
i. 2 11/16 in.
j. 2 1/4 in.
k. 2 1/8 in.
l. 2 3/16 in.

6\. a. either 3 inches or 3 1/4 inches; 3 1/8 in.; 3 2/16 in.
b. 3/4 in.; either 5/8 or 6/8 in.; 11/16 in.
c. 2 3/4 in. on all rulers
d. 1 in; 1 in; 15/16 in.
e. 1 1/4 in.; 1 3/8 in.; 1 5/16 in.
f. 1 3/4 in. on all rulers.

7\. Answers will vary. Please check the student's work.

Measuring in Inches, cont.

8. The sides measure: 1 1/2 in., 3 15/16 in., 1 11/16 in., and 3 1/4 in. The perimeter is 10 3/8 in.

9. 49 3/4 inches

10. a.

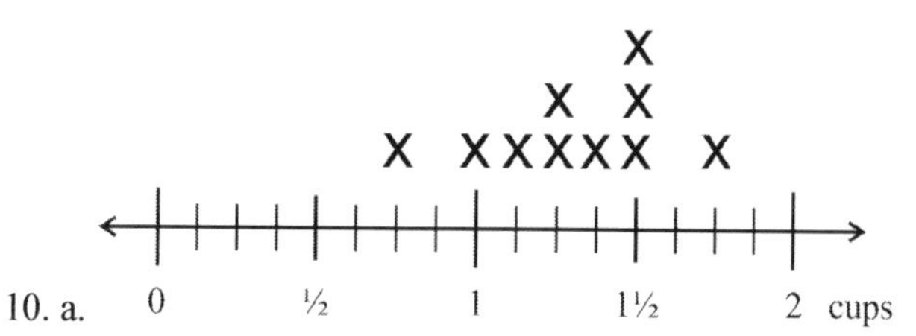

b. 3/4 C + 3/4 C + 3/4 C = 2 1/4 C

c. 1 3/4 C + 1 3/4 C + 1 3/4 C = 5 1/4 C

11. a.

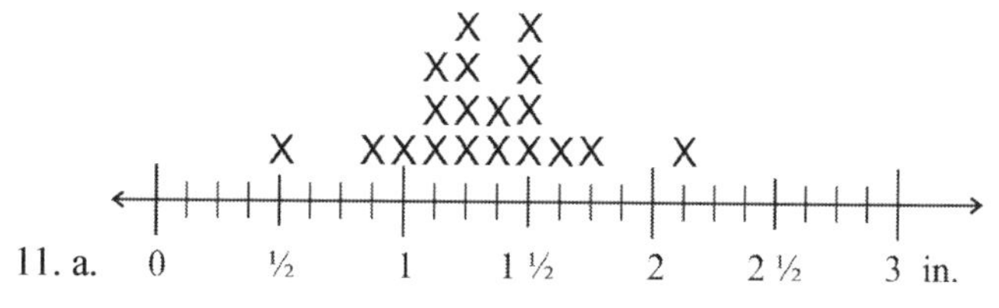

b. There are two modes: 1 1/4 in. and 1 1/2 in.

c. 2 1/8 in. + 1 3/4 in. + 1 5/8 in. + 1 1/2 in. + 1 1/2 in. =

2 1/8 in. + 1 6/8 in. + 1 5/8 in. + 3 in. = 8 4/8 in. = 8 1/2 in.

12. Answers will vary. Check the students' work.

Mixed Review, p. 92

1.

a. 452,912,980 Place: ten thousands place Value: 10,000 (ten thousand)	b. 6,219,455,221 Place: billions place Value: 6,000,000,000 (six billion)

2. 60 × 60 = 3,600 seconds in an hour; 3,600 × 24 = 86,400 seconds in a day

3. (52 − 12) × 5 × 5 = 1,000. They do 1,000 hours of schoolwork in a year.

4. Estimate: 2 × 7 = 14. Exact answer: 14.348

5. a. 0.034; 0.021 b. 9.8; 46,700 c. 0.019; 300

6.

a. 5,070 g = 5.07 kg 2.5 kg = 2,500 g	b. 0.6 L = 600 ml 10,500 ml = 10.5 L	c. 0.06 km = 60 m 2,600 m = 2.6 km

7. a. 82.50 ÷ 0.06 = 825 ÷ 0.6 = 8,250 ÷ 6 = 1,375 b. 48.302 ÷ 0.2 = 483.02 ÷ 2 = 241.51

8. 3 × 2.40 − 0.15 + 0.30 = $7.35 for the three cups of yogurt.

9. $6.29 × 3 = $18.87 is the price of Shirt B.

Mixed Review, cont.

10. Multiplying as if there was no decimal point, I get 1,000 × _7_. That equals _7,000_.
Then, since my answer has to have thousandths, it needs _3_ decimal digits.
So, the final answer is _7.000 or 7_.

11. a.

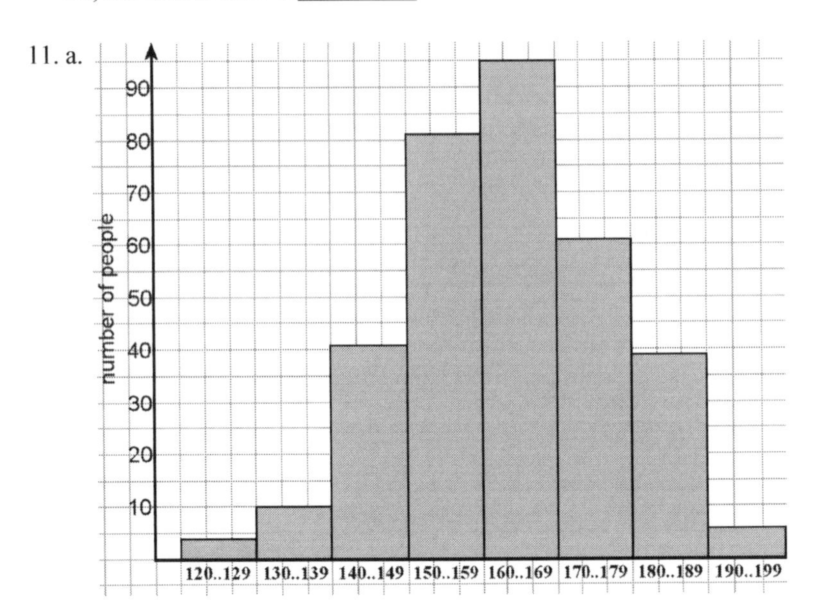

b. 14 people c. 45 people d. About 82 + 41 + 10 + 4 = 137 children, about 95 + 61 + 39 + 6 = 201 adults
e. It could be a group of people that were at the swimming pool at 5 pm on a certain Tuesday because there were both children and adults.

Review, p. 95

1. a. 19/2 b. 61/11 c. 58/7 d. 506/100

2. a. 4 1/10 b. 6 1/3 c. 3 1/9 d. 2 8/12

3. 23/6 = 3 5/6

4. a. 5 5/8 b. 6 12/20 c. 5 4/15

5. a. 15/21 + 7/21 = 22/21 = 1 1/21 b. 9/30 + 10/30 = 19/30
c. 2 9/7 − 1 6/7 = 1 3/7 d. 2 16/20 + 3 5/20 = 5 21/20 = 6 1/20

6. a. < b. < c. = d. < e. < f. > g. < h. <

7. From the first piece, she has left: 5 1/2 ft − 3 1/8 ft = 5 4/8 ft − 3 1/8 ft = 2 3/8 ft.
From the second piece, she has left: 4 1/4 ft − 3 1/8 ft = 4 2/8 ft − 3 1/8 ft = 1 1/8 ft.
Combined, those two pieces are 2 3/8 ft + 1 1/8 ft = 3 4/8 ft = 3 1/2 ft.

8. 1 − 32/100 − 42/100 − 2/10 = 1 − 32/100 − 42/100 − 20/100 = 6/100. So, 6/100 of the land is resting.

9. One-fifth of $35 is $7, so the discounted price would be $28. And 2/11 of $33 is $6, so its discounted price would be $27. So, 2/11 off of the $33-book is the better deal.
If both books cost $50, then 1/5 off of it would be the better buy. This is because 1/5 is more than 2/11.

Chapter 7: Fractions: Multiply and Divide

Simplifying Fractions, p. 100

1.

a. The parts were joined together in fours.

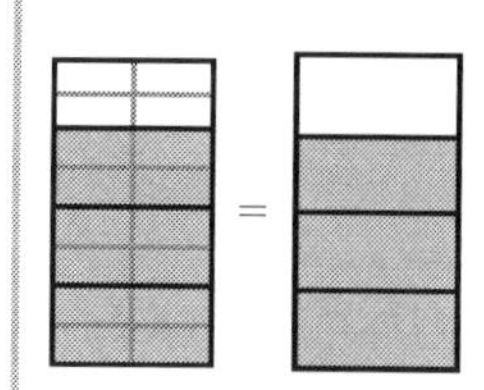

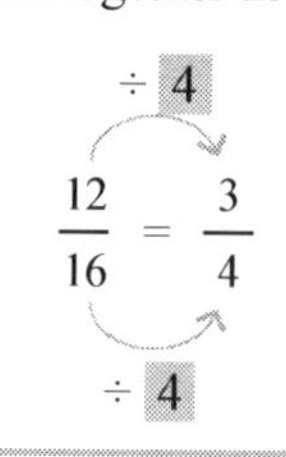

$\frac{12}{16} = \frac{3}{4}$ (÷ 4)

b. The parts were joined together in eights.

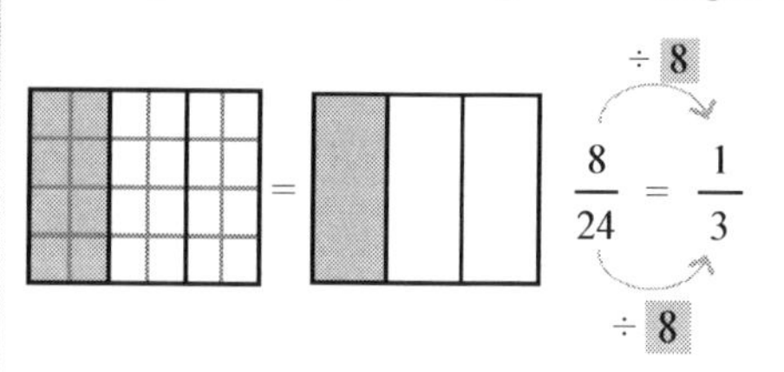

$\frac{8}{24} = \frac{1}{3}$ (÷ 8)

2.

a. The slices were joined together in threes.

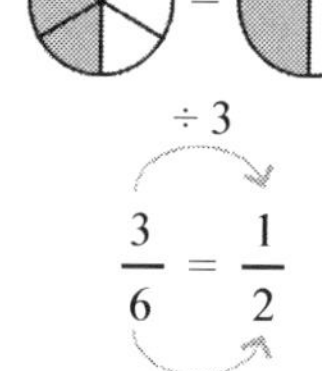

$\frac{3}{6} = \frac{1}{2}$ (÷ 3)

b. The slices were joined together in fours.

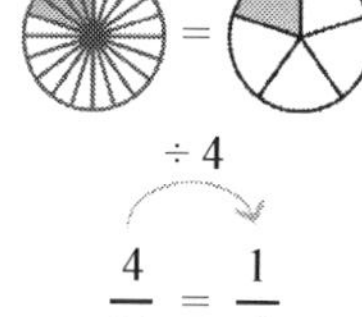

$\frac{4}{20} = \frac{1}{5}$ (÷ 4)

c. The slices were joined together in threes.

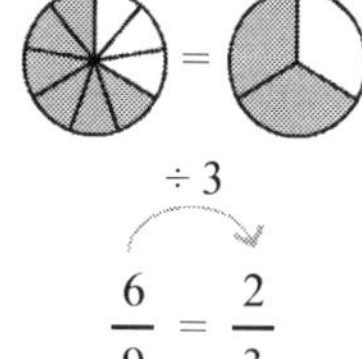

$\frac{6}{9} = \frac{2}{3}$ (÷ 3)

d. The slices were joined together in fours.

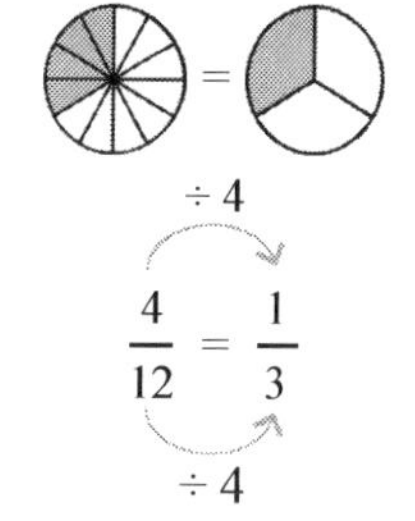

$\frac{4}{12} = \frac{1}{3}$ (÷ 4)

e. The slices were joined together in threes.

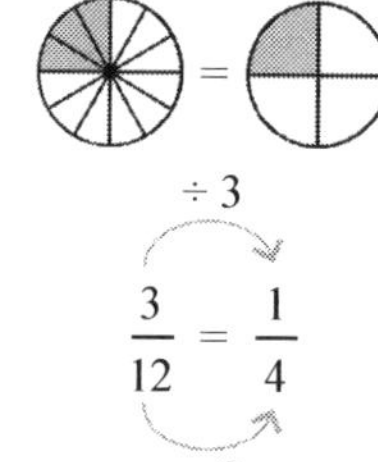

$\frac{3}{12} = \frac{1}{4}$ (÷ 3)

f. The slices were joined together in fives.

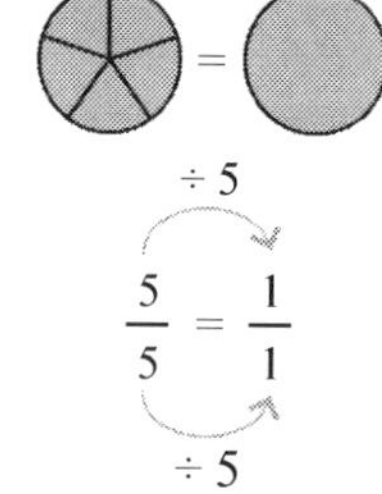

$\frac{5}{5} = \frac{1}{1}$ (÷ 5)

g. The slices were joined together in twos.

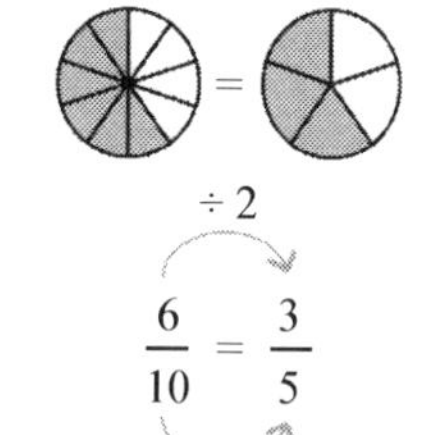

$\frac{6}{10} = \frac{3}{5}$ (÷ 2)

h. The slices were joined together in twos.

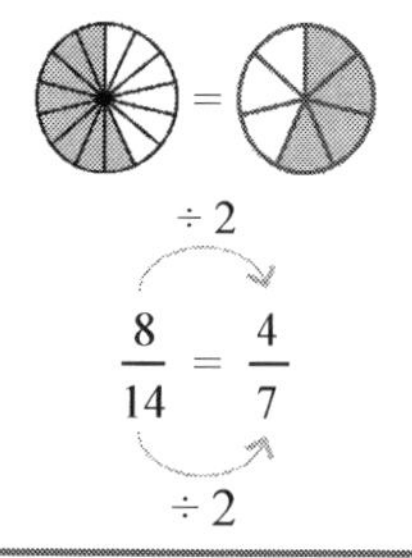

$\frac{8}{14} = \frac{4}{7}$ (÷ 2)

3.

a. $\frac{2}{6} = \frac{1}{3}$

b.

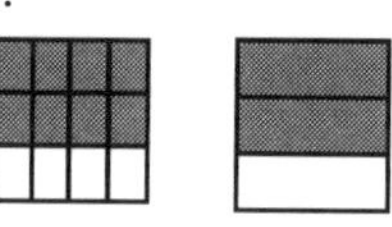

$\frac{8}{12} = \frac{2}{3}$

c.

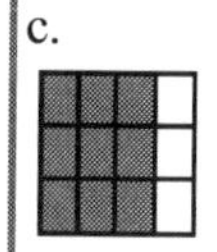

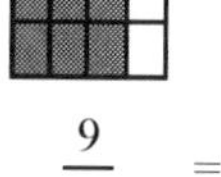

$\frac{9}{12} = \frac{3}{4}$

d. $\frac{6}{18} = \frac{1}{3}$

e.

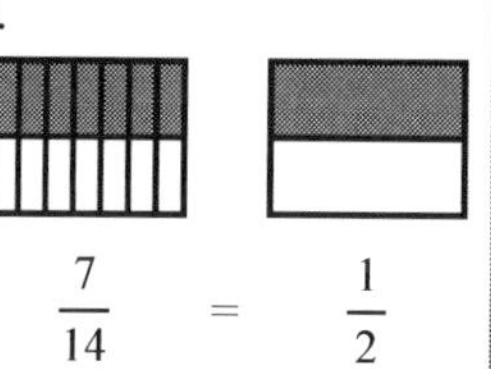

$\frac{7}{14} = \frac{1}{2}$

f.

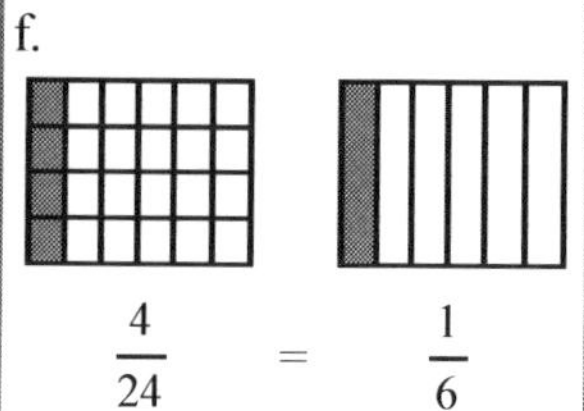

$\frac{4}{24} = \frac{1}{6}$

Simplifying Fractions 1, cont.

4.

a. $\frac{6}{16} = \frac{3}{8}$ (÷ 2)	b. $\frac{15}{25} = \frac{3}{5}$ (÷ 5)	c. $\frac{3}{9} = \frac{1}{3}$ (÷ 3)	d. $\frac{4}{8} = \frac{1}{2}$ (÷ 4)	e. $\frac{16}{24} = \frac{2}{3}$ (÷ 8)
f. $\frac{12}{20} = \frac{3}{5}$	g. $\frac{24}{32} = \frac{3}{4}$	h. $\frac{3}{15} = \frac{1}{5}$	i. $\frac{15}{18} = \frac{5}{6}$	j. $\frac{16}{20} = \frac{4}{5}$

5. b. 5 1/9 c. 7 1/4 d. 3 2/7

6. a. cannot simplify b. 1/3 c. cannot simplify d. cannot simplify
   e. 1/2 f. 1/2 g. cannot simplify h. cannot simplify
   i. cannot simplify j. 1/7 k .cannot simplify l. 2/11

7.

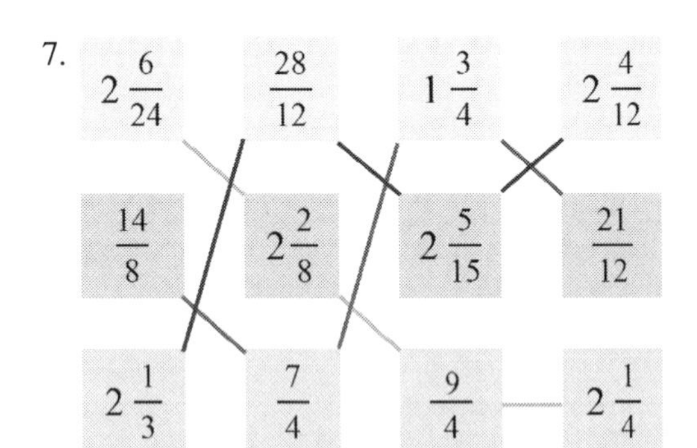

8. What part of the total time is the warm-up time? 1/6
   What part of the total time is the actual running practice time? 2/3

9. Answers will vary.

Simplifying Fractions 2, p. 104

1.

a. You could simplify in one step if you divided by 40. $\frac{40}{120} = \frac{4}{12} = \frac{1}{3}$ (÷ 10, ÷ 4)	b. You could simplify in one step if you divided by 15. $\frac{75}{105} = \frac{15}{21} = \frac{5}{7}$ (÷ 5, ÷ 3)
c. You could simplify in one step if you divided by 6. $\frac{30}{96} = \frac{10}{32} = \frac{5}{16}$ (÷ 3, ÷ 2)	d. You could simplify in one step if you divided by 14. $\frac{42}{98} = \frac{21}{49} = \frac{3}{7}$ (÷ 2, ÷ 7)

2. a. 1/3 b. 1/4 c. 3/4 d. 21/25 e. 3/10 f. 3/8 g. 5/14 h. 3/10 i. 5 1/3 j. 5 3/4 k. 3 2/7 l. 7 3/5

3. b. 3/2 = 1 1/2 c. 16/9 = 1 7/9 d. 8/7 = 1 1/7 e. 4/3 = 1 1/3 f. 11/6 = 1 5/6

Simplifying Fractions 2, cont.

4. Who got it right? Nancy and Jerry got it right. Who didn't? Mark did not.
   Why? Because Mark did not reduce it to the lowest terms. His answer was 8/10, which can still be simplified into 4/5.

5. a. cannot simplify b. 7/11 c. cannot simplify d. 4/9 e. 2 3/4
   f. 2 5/11 (cannot simplify) g. cannot simplify h. 1 3/5 i. 4 2/5

6. a. 5/6 + 1 4/6 = 2 3/6 = 2 1/2
   b. 7/12 + 9/12 = 16/12 = 1 4/12 = 1 1/3
   c. 21/14 + 12/14 = 33/14 = 2 5/14
   d. 9/10 − 1/15 = 25/30 = 5/6
   e. 7/8 − 5/6 = 2/48 = 1/24
   f. 15/8 − 3/10 = 63/40 = 1 23/40
   g. 3 3/4 − 1 5/6 = 23/12 = 1 11/12
   h. 5 5/9 + 3 7/12 = 329/36 = 9 5/36

7. a. 3/4 b. 600 pixels

8.

	$\frac{3}{4}$	$\frac{2}{5}$	$\frac{1}{2}$	$\frac{2}{7}$		$\frac{1}{4}$	$\frac{3}{5}$	$\frac{1}{2}$		$\frac{1}{4}$	$\frac{2}{3}$	$\frac{1}{6}$	$\frac{1}{4}$	$\frac{2}{7}$	$\frac{1}{3}$		$\frac{1}{3}$	$\frac{3}{4}$	$\frac{5}{6}$	$\frac{3}{10}$	$\frac{3}{10}$	$\frac{1}{2}$	$\frac{3}{7}$	
Because	T	H	E	Y		A	R	E		A	L	W	A	Y	S		S	T	U	F	F	E	D	.

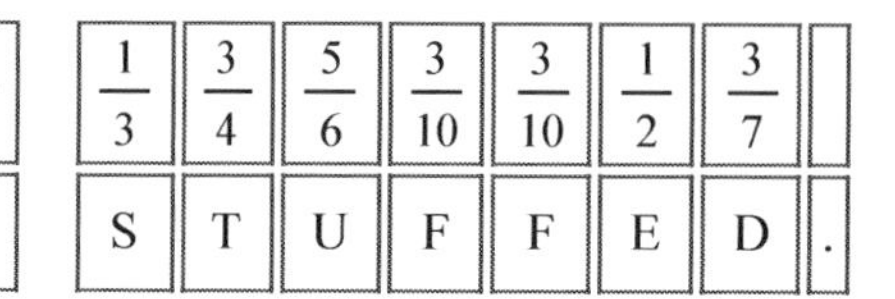

Multiply Fractions by Whole Numbers, p. 108

1.

a. $4 \times \frac{7}{9} = \frac{28}{9} = 3\frac{1}{9}$

b. $3 \times \frac{5}{8} = \frac{15}{8} = 1\frac{7}{8}$

c. $5 \times \frac{11}{12} = \frac{55}{12} = 4\frac{7}{12}$

d. $6 \times \frac{7}{10} = \frac{42}{10} = 4\frac{2}{10} = 4\frac{1}{5}$

2.

a. $2\frac{4}{5} = 2 \times 1\frac{2}{5}$ b. $\frac{25}{9} = 5 \times \frac{5}{9}$ c. $2\frac{2}{8} = 3 \times \frac{6}{8}$

3. 4 × (3/8) = 12/8 = 1 4/8 = 1 1/2 liters

4. See the tripled recipe on the right.

Brownies

2 1/4 cup butter
4 1/2 cups brown sugar
12 eggs
3 3/4 cups cocoa powder
1 1/2 cup flour
6 tsp vanilla

5. b. 28/10 = 2 4/5 c. 22/20 = 1 2/20 = 1 1/10 d. 18/15 = 6/5 = 1 1/5
   e. 30/6 = 5 f. 42/100 = 21/50 g. 16/12 = 4/3 = 1 1/3
   h. 70/100 = 7/10 i. 90/20 = 18/4 = 4 1/2 j. 49/15 = 3 4/15

6. a. Most students used between 1/2 and 1 1/4 hours for housework and chores.
   b. Multiply the average by the number of students to get the total:
   20 × (7/8) = 140/8 = 17 4/8 = 17 1/2 hours

7. a. 14 lb b. 80 km

8. 160 cm and 240 cm, or 1.6 m and 2.4 m

Multiplying Fractions by Whole Numbers, cont.

9. a. Since 1/3 of $81 is $27, Janet got $54 and Sandy got $27.

   b. We would need to use long division, and divide $80.00 by 3. The division will not end, so the answer needs to be rounded. Now, Sandy would get $26.67 and Janet would get the rest, or $53.33.

```
      2 6.6 6 6
   3 )8 0.0 0 0
     - 6
       2 0
      -1 8
         2 0
       - 1 8
           2 0
          -1 8
             2 0
```

10. a. Three-fourths of 5 in. is calculated as 3/4 × 5 in. = 15/4 in. = 3 3/4 in. Three-fourths of 4 in. is 3 in. His second rectangle was 3 3/4 in. by 3 in.

    b. Check students' rectangles.

11.

a. A two-fifth part of 10	**10 copies of 2/5**
$\frac{2}{5} \times 10$ means a two-fifth part of 10, which is equal to 4.	$10 \times \frac{2}{5}$ means 10 copies of 2/5, which is equal to 4.

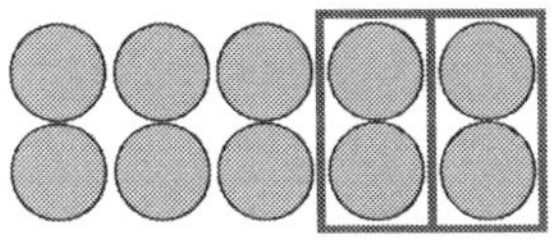

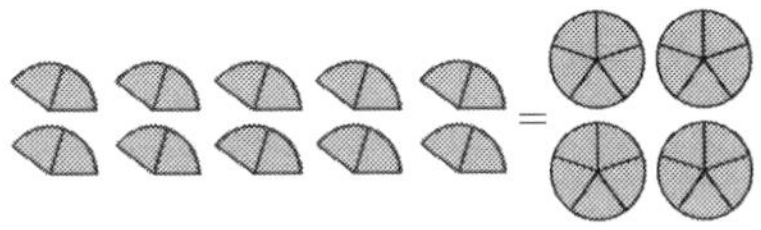

b. A third part of 5	**5 copies of 1/3**
$\frac{1}{3} \times 5$ means a third part of 5, which is equal to 1 2/3.	$5 \times \frac{1}{3}$ means 5 copies of 1/3, which is equal to 1 2/3.

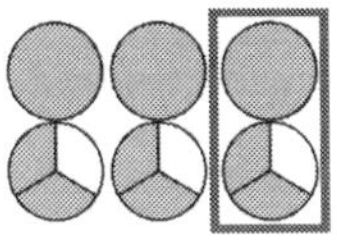

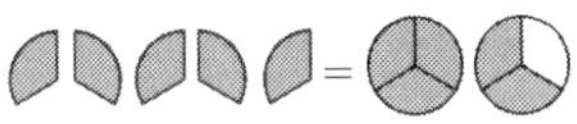

c. three-fourths of 7	**7 copies of 3/4**
$\frac{3}{4} \times 7$ means three-fourths of 7, which is equal to 3 9/4 = 5 1/4.	$7 \times \frac{3}{4}$ means 7 copies of 3/4, which is equal to 5 1/4.

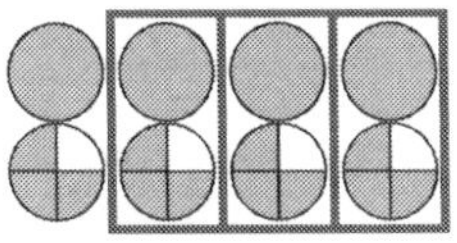

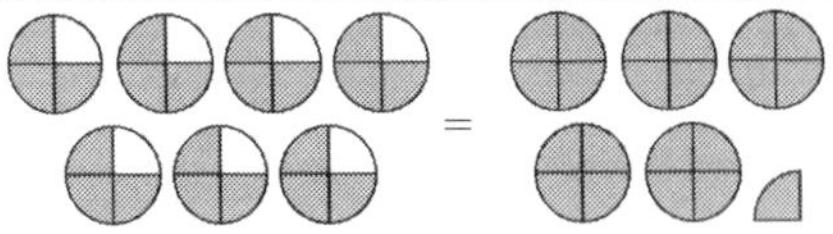

Multiplying Fractions by Fractions, p. 112

1.

a. $\frac{1}{2} \times \frac{1}{4} = \frac{1}{8}$	b. $\frac{1}{2} \times \frac{1}{2} = \frac{1}{4}$	c. $\frac{1}{2} \times \frac{1}{5} = \frac{1}{10}$
d. $\frac{1}{3} \times \frac{1}{2} = \frac{1}{6}$	e. $\frac{1}{3} \times \frac{1}{3} = \frac{1}{9}$	f. $\frac{1}{3} \times \frac{1}{4} = \frac{1}{12}$
g. $\frac{1}{4} \times \frac{1}{2} = \frac{1}{8}$	h. $\frac{1}{4} \times \frac{1}{3} = \frac{1}{12}$	i. $\frac{1}{4} \times \frac{1}{4} = \frac{1}{16}$
Did you notice a shortcut? You can simply multiply the denominators. $\frac{1}{5} \times \frac{1}{6} = \frac{1}{30}$		

2. a. 1/18 b. 1/39 c. 1/100

3.

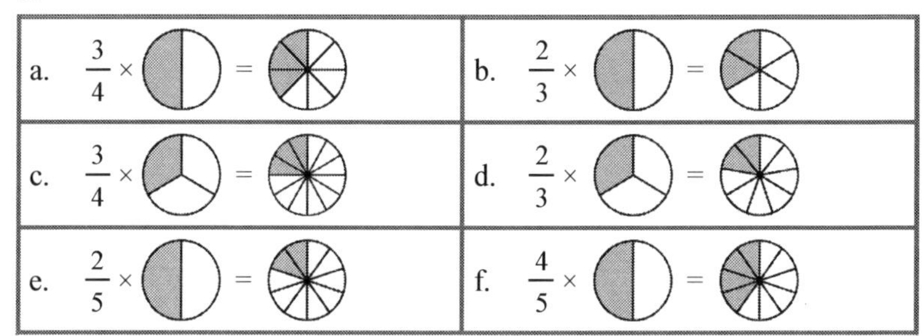

4.

a. $\frac{2}{3} \times \frac{1}{8} =$ First find 1/3 of 1/8, then multiply the result by 2. $\frac{1}{3} \times \frac{1}{8} = \frac{1}{24}$ and $\frac{1}{24} \times 2 = \frac{2}{24} = \frac{1}{12}$	b. $\frac{3}{4} \times \frac{1}{10} =$ First find 1/4 of 1/10, then multiply the result by 3. $\frac{1}{4} \times \frac{1}{10} = \frac{1}{40}$ and $\frac{1}{40} \times 3 = \frac{3}{40}$
c. $\frac{3}{5} \times \frac{1}{6} =$ First find 1/5 of 1/6, then multiply the result by 3. $\frac{1}{5} \times \frac{1}{6} = \frac{1}{30}$ and $\frac{1}{30} \times 3 = \frac{3}{30} = \frac{1}{10}$	d. $\frac{5}{6} \times \frac{1}{9} =$ First find 1/6 of 1/9, then multiply the result by 5. $\frac{1}{6} \times \frac{1}{9} = \frac{1}{54}$ and $\frac{1}{54} \times 5 = \frac{5}{54}$
e. $\frac{2}{3} \times \frac{1}{7} =$ (First find 1/3 of 1/7, then multiply the result by 2.) $\frac{1}{3} \times \frac{1}{7} = \frac{1}{21}$ and $\frac{1}{21} \times 2 = \frac{2}{21}$	f. $\frac{3}{8} \times \frac{1}{4} =$ (First find 1/8 of 1/4, then multiply the result by 3.) $\frac{1}{8} \times \frac{1}{4} = \frac{1}{32}$ and $\frac{1}{32} \times 3 = \frac{3}{32}$

5. a. 2/27 b. 11/72 c. 1/13 d. 6 e. 4/21 f. 7 1/7

6. a. 21/32 b. 1 3/25 c. 9/25 d. 2/15 e. 1/14 f. 3/4 g. 11/12 h. 1/15

7. a. 2/12 or 1/6. Multiply (2/3) × (1/4) to get the answer.
b. 1/12, because 1/4 of the pizza was left, which is equal to 3/12, and Marie ate 2/12.

8. a. 3/8

b. At first, Theresa painted 5/8 of the room.

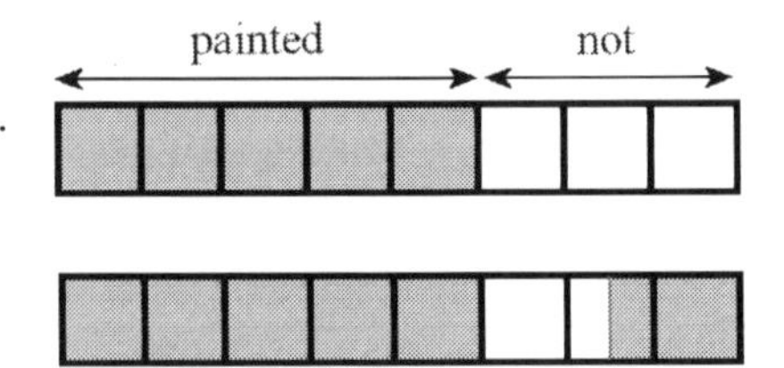

Later, she painted 1/2 of what was left.

Now there is still 1/8 and half of 1/8 of the room left to paint.
Since half of 1/8 is 1/16, there is 3/16 of the room still left to paint

9. a. 1/3

b.

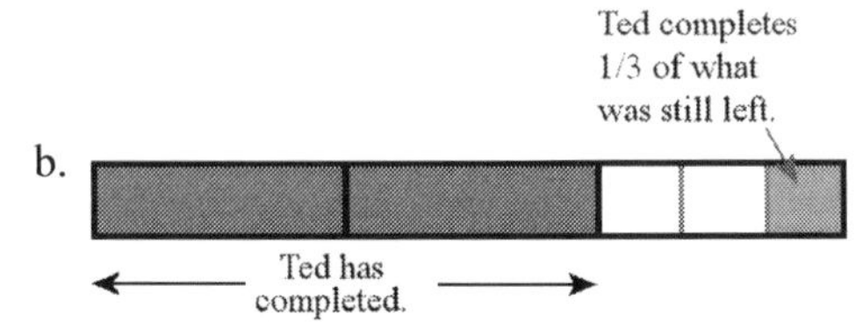

Now, 2/9 of the job is still not done, and 7/9 of it is completed.

10. See the recipe on the right.

Carob Brownies (1/3 recipe)

1 cup sweetened carob chips
2 2/3 tablespoons Extra virgin olive oil
1 (small) egg
1/6 cup honey
1/3 teaspoon vanilla
1/4 cup whole wheat flour
1/4 teaspoon baking powder
1/3 cup walnuts or other nuts

11. a. Thirty guests will drink 15 + 30 = 45 servings, so Alison needs to make the recipe 9 times. That means she needs 9 × 1/4 = 2 1/4 cups of coffee.

b. Fifty guests will drink 25 + 50 = 75 servings, so Alison needs to make the recipe 15 times. That means she needs 15 × 1/4 = 3 3/4 cups of coffee.

c. Eighty guests will drink 40 + 80 = 120 servings, so Alison needs to make the recipe 24 times. That means she needs 24 × 1/4 = 6 cups of coffee.

Puzzle corner: a. 1/6 b. 5/4 c. 1/6 d. 3/4

Fraction Multiplication and Area, p. 117

1.

a.

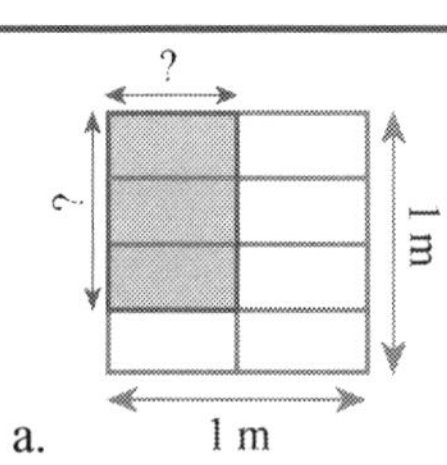

Side lengths: $\frac{1}{2}$ m and $\frac{3}{4}$ m

Area (from the picture): $\frac{3}{8}$ m^2

b.

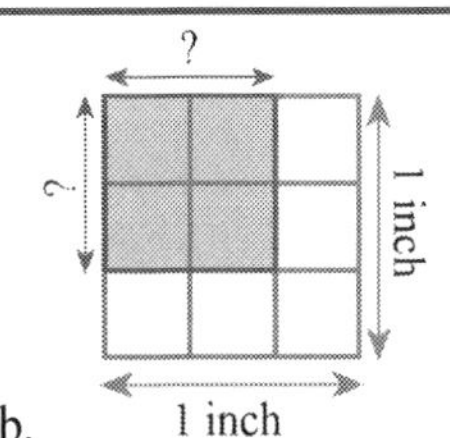

Side lengths: $\frac{2}{3}$ in. and $\frac{2}{3}$ in.

Area (from the picture): $\frac{4}{9}$ in^2

2.

a.

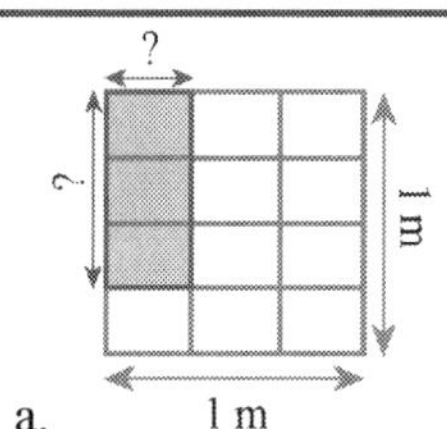

Side lengths: $\frac{1}{3}$ m and $\frac{3}{4}$ m

Area (by multiplication):

$\frac{1}{3}$ m × $\frac{3}{4}$ m = $\frac{3}{12}$ m^2

b.

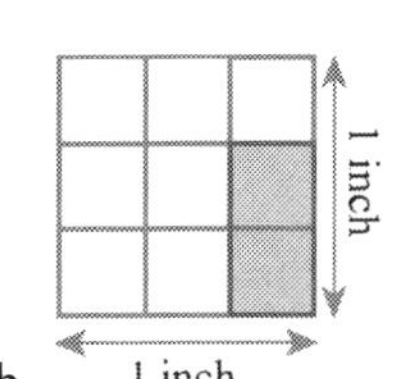

Side lengths: $\frac{1}{3}$ in. and $\frac{2}{3}$ in.

Area (by multiplication):

$\frac{1}{3}$ in. × $\frac{2}{3}$ in. = $\frac{2}{9}$ in^2

c.

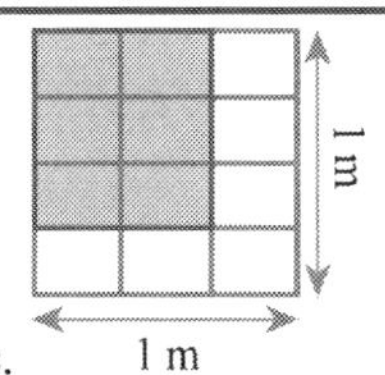

Side lengths: $\frac{2}{3}$ m and $\frac{3}{4}$ m

Area (by multiplication):

$\frac{2}{3}$ m × $\frac{3}{4}$ m = $\frac{6}{12}$ m^2

d.

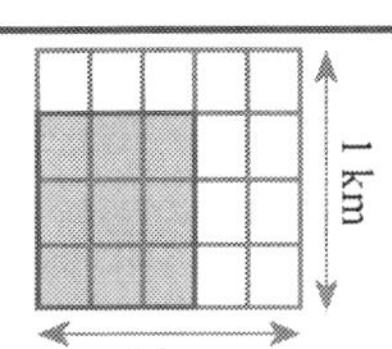

Side lengths: $\frac{3}{5}$ km and $\frac{3}{4}$ km

Area (by multiplication):

$\frac{3}{5}$ km × $\frac{3}{4}$ km = $\frac{9}{20}$ km^2

3. The coloring may vary, but students' pictures should either look like these, or be rotated versions of these.

a.  $\frac{1}{4}$ m × $\frac{1}{2}$ m = $\frac{1}{8}$ m^2

b. 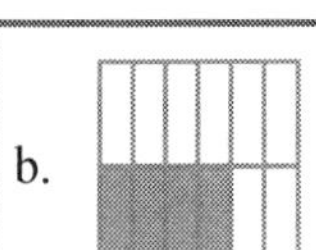 $\frac{1}{2}$ in. × $\frac{4}{6}$ in. = $\frac{4}{12}$ in^2

c. $\frac{3}{4}$ ft × $\frac{2}{7}$ ft = $\frac{6}{28}$ ft^2

d. $\frac{3}{5}$ km × $\frac{5}{6}$ km = $\frac{15}{30}$ km^2

Fraction Multiplication and Area, cont.

4. In each problem, the factors may also be written in the other order.

a. 1/3 m

1/3 m

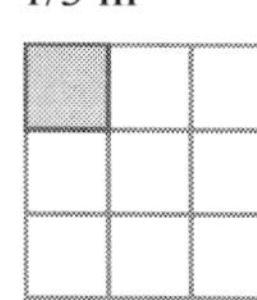

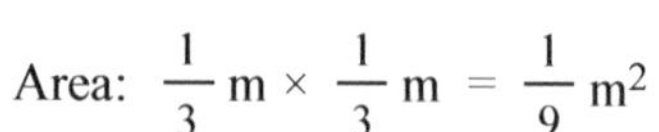

Area: $\frac{1}{3}$ m × $\frac{1}{3}$ m = $\frac{1}{9}$ m^2

b. 1/5 m

1/3 m

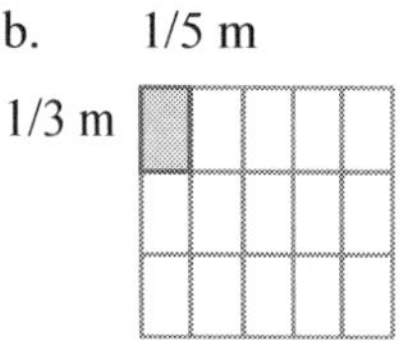

Area: $\frac{1}{3}$ m × $\frac{1}{5}$ m = $\frac{1}{15}$ m^2

c. 1/5 m

1/2 m

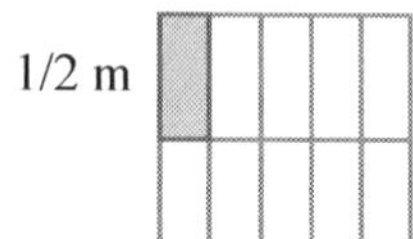

Area: $\frac{1}{5}$ m × $\frac{1}{2}$ m = $\frac{1}{10}$ m^2

d. 1/4 m

1/4 m

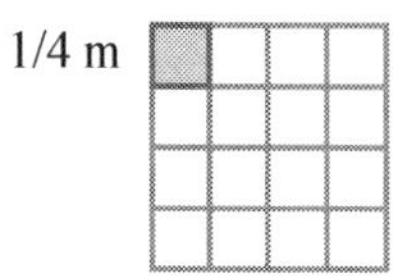

Area: $\frac{1}{4}$ m × $\frac{1}{4}$ m = $\frac{1}{16}$ m^2

5. In each problem, the factors may also be written in the other order.

a. 3/4 m

1/2 m

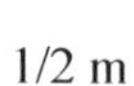

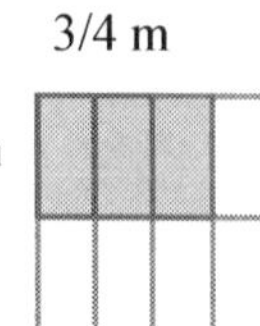

Area: $\frac{3}{4}$ m × $\frac{1}{2}$ m = $\frac{3}{8}$ m^2

b. 2/5 m

3/4 m

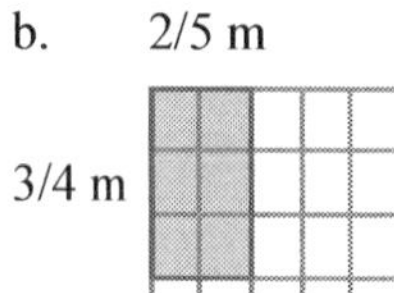

Area: $\frac{2}{5}$ m × $\frac{3}{4}$ m = $\frac{6}{20}$ m^2

c. 2/3 m

2/3 m

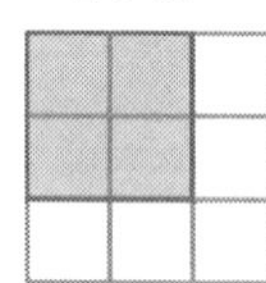

Area: $\frac{2}{3}$ m × $\frac{2}{3}$ m = $\frac{4}{9}$ m^2

d. 3/5 m

1/2 m

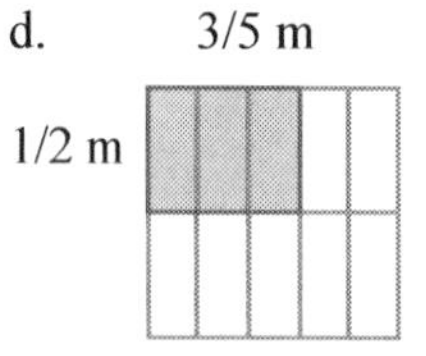

Area: $\frac{3}{5}$ m × $\frac{1}{2}$ m = $\frac{3}{10}$ m^2

e. 3/4 m

3/4 m

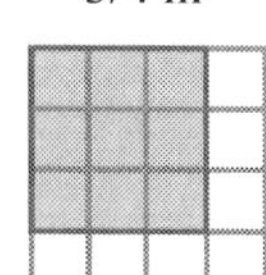

Area: $\frac{3}{4}$ m × $\frac{3}{4}$ m = $\frac{9}{16}$ m^2

f. 5/6 m

1/2 m

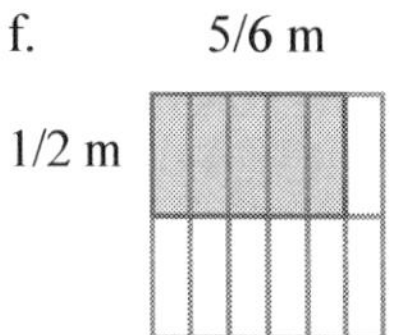

Area: $\frac{5}{6}$ m × $\frac{1}{2}$ m = $\frac{5}{12}$ m^2

6. In each problem, the factors may also be written in the other order.

a. $\frac{3}{5} \times \frac{3}{4} = \frac{9}{20}$ b. $\frac{3}{5} \times \frac{2}{3} = \frac{6}{15}$

c. $\frac{2}{3} \times \frac{2}{5} = \frac{4}{15}$ d. $\frac{2}{2} \times \frac{2}{4} = \frac{4}{8}$

7. a. 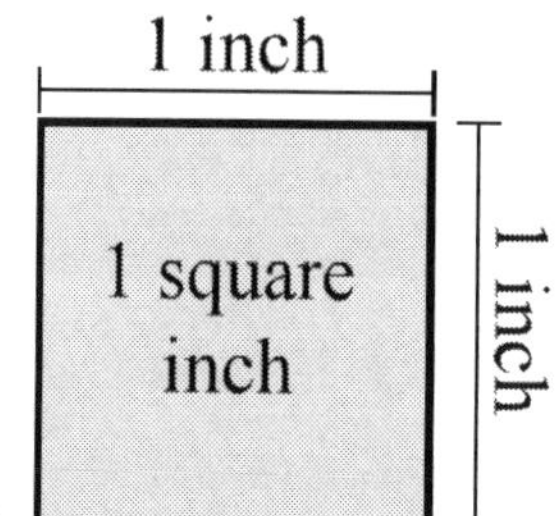

Its area is 1 square inch.

b. Check students' work.
c. The area is (3/4 in.) × (5/8 in.) = 15/32 square inches.

8. a. 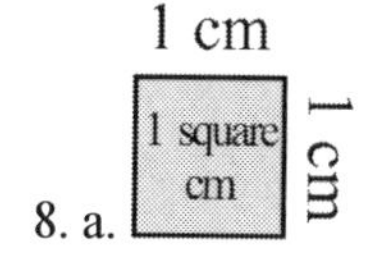

Check students' work.

b. 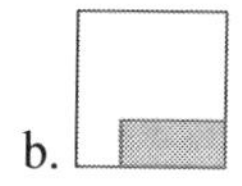

c. Using fractions: (3/10 cm) × (7/10 cm) = 21/100 cm^2.
Using decimals: 0.3 cm × 0.7 cm = 0.21 cm^2.

9. a. The area is 3 km × (1/2 km) = 1 ½ km^2.

b. The area is (5/8 mi.) × (3/4 mi.) = 15/32 mi^2.

10. a. (3 3/5 km) × ($3 ½ per km) = (18/5) × (7/2) = 126/10 = 12 6/10 dollars.

b. 3.6 km × $3.5 per km = 12.6 dollars = $12.60.

11. a. The area of one stamp is (7/8 in.) × (3/4 in.) = 21/32 square inches.
The area of six stamps is 6 × (21/32 in^2) = 126/32 in^2 = 63/16 in^2 = 3 15/16 in^2.

b. The area of the envelope is 40 square inches. The stamps cover about 1/10 of it (the area of the stamps is very close to 4 square inches).

Puzzle corner: The area of the square is (7/8 in.) × (7/8 in.) = 49/64 in^2. The area of the rectangle is 3/4 in^2 = 48/64 in^2. The 7/8-in. by 7/8-in. square has a larger area than the rectangle. The difference is 1/64 square mile.

Simplifying Before Multiplying 1, p. 123

1.

a. $\frac{\overset{7}{\cancel{14}}}{\underset{8}{\cancel{16}}} = \frac{7}{8}$	b. $\frac{\overset{11}{\cancel{33}}}{\underset{9}{\cancel{27}}} = \frac{11}{9}$	c. $\frac{\overset{6}{\cancel{12}}}{\underset{13}{\cancel{26}}} = \frac{6}{13}$	d. $\frac{\overset{3}{\cancel{9}}}{\underset{11}{\cancel{33}}} = \frac{3}{11}$

2. Students do not need to write the intermediate multiplication step. For clarity, it is shown here in its entirety.

a. $\frac{\overset{3}{\cancel{6}}}{\underset{5}{\cancel{10}}} \times \frac{\overset{1}{\cancel{2}}}{\underset{7}{\cancel{14}}} = \frac{3 \times 1}{5 \times 7} = \frac{3}{35}$	b. $\frac{\overset{1}{\cancel{2}}}{\underset{2}{\cancel{4}}} \times \frac{\overset{1}{\cancel{3}}}{\underset{5}{\cancel{15}}} = \frac{1 \times 1}{2 \times 5} = \frac{1}{10}$
c. $\frac{\overset{2}{\cancel{8}}}{\underset{3}{\cancel{12}}} \times \frac{1}{2} = \frac{2 \times 1}{3 \times 2} = \frac{2}{6} = \frac{1}{3}$	d. $\frac{\overset{1}{\cancel{8}}}{\underset{4}{\cancel{32}}} \times \frac{\overset{2}{\cancel{14}}}{\underset{3}{\cancel{21}}} = \frac{1 \times 2}{4 \times 3} = \frac{2}{12} = \frac{1}{6}$
e. $\frac{\overset{2}{\cancel{6}}}{\underset{5}{\cancel{15}}} \times \frac{\overset{2}{\cancel{6}}}{\underset{3}{\cancel{9}}} = \frac{2 \times 2}{5 \times 3} = \frac{4}{15}$	f. $\frac{\overset{3}{\cancel{27}}}{\underset{5}{\cancel{45}}} \times \frac{\overset{3}{\cancel{21}}}{\underset{7}{\cancel{49}}} = \frac{3 \times 3}{5 \times 7} = \frac{9}{35}$

3. Students do not need to write the intermediate multiplication step. For clarity, it is shown here in its entirety.

a. $\frac{8}{\underset{3}{\cancel{9}}} \times \frac{\overset{2}{\cancel{6}}}{11} = \frac{8 \times 2}{3 \times 11} = \frac{16}{33}$	b. $\frac{3}{\underset{5}{\cancel{10}}} \times \frac{\overset{1}{\cancel{2}}}{5} = \frac{3 \times 1}{5 \times 5} = \frac{3}{25}$	c. $\frac{\overset{1}{\cancel{4}}}{7} \times \frac{1}{\underset{3}{\cancel{12}}} = \frac{1 \times 1}{7 \times 3} = \frac{1}{21}$
d. $\frac{\overset{1}{\cancel{7}}}{4} \times \frac{3}{\underset{3}{\cancel{21}}} = \frac{1 \times 3}{4 \times 3} = \frac{3}{12} = \frac{1}{4}$	e. $\frac{3}{\underset{2}{\cancel{16}}} \times \frac{\overset{1}{\cancel{8}}}{5} = \frac{3 \times 1}{2 \times 5} = \frac{3}{10}$	f. $\frac{3}{\underset{2}{\cancel{8}}} \times \frac{\overset{3}{\cancel{12}}}{11} = \frac{3 \times 3}{2 \times 11} = \frac{9}{22}$

4. Students do not need to write the intermediate multiplication step. For clarity, it is shown here in its entirety.

a. $\frac{\overset{1}{\cancel{7}}}{\underset{4}{\cancel{8}}} \times \frac{\overset{1}{\cancel{2}}}{\underset{1}{\cancel{7}}} = \frac{1 \times 1}{4 \times 1} = \frac{1}{4}$	b. $\frac{\overset{1}{\cancel{3}}}{\underset{1}{\cancel{5}}} \times \frac{\overset{1}{\cancel{5}}}{\underset{2}{\cancel{6}}} = \frac{1 \times 1}{1 \times 2} = \frac{1}{2}$	c. $\frac{\overset{1}{\cancel{5}}}{\underset{3}{\cancel{12}}} \times \frac{\overset{1}{\cancel{4}}}{\underset{2}{\cancel{10}}} = \frac{1 \times 1}{3 \times 2} = \frac{1}{6}$
d. $\frac{\overset{1}{\cancel{9}}}{\underset{5}{\cancel{15}}} \times \frac{\overset{1}{\cancel{3}}}{\underset{2}{\cancel{18}}} = \frac{1 \times 1}{5 \times 2} = \frac{1}{10}$	e. $\frac{\overset{2}{\cancel{8}}}{11} \times \frac{3}{\underset{1}{\cancel{4}}} = \frac{2 \times 3}{11 \times 1} = \frac{6}{11}$	f. $\frac{\overset{4}{\cancel{12}}}{\underset{25}{\cancel{100}}} \times \frac{\overset{1}{\cancel{4}}}{\underset{5}{\cancel{15}}} = \frac{4 \times 1}{25 \times 5} = \frac{4}{125}$

Simplifying Before Multiplying, cont.

5.

a. $\frac{82}{\cancel{77}_1} \times \cancel{77}^1 = 82$	b. $\cancel{13}^1 \times \frac{49}{\cancel{13}_1} = 49$	c. $\frac{\cancel{14}^1 \times 16}{\cancel{14}_1} = 16$
d. $\frac{5}{\cancel{6}_1} \times \cancel{24}^4 = 20$	e. $\cancel{54}^6 \times \frac{2}{\cancel{9}_1} = 12$	f. $\frac{\cancel{16}^2 \times 5}{\cancel{8}_1} = 10$

6. A stack of eight is 8 × (3/8 in.) = 3 in. tall. A stack of twenty is 20 × (3/8 in.) = 5 × (3/2 in.) = 15/2 in. = 7 1/2 in. tall.

7. 52 × (3/4 kg) = 13 × (3 kg) = 39 kg.

8. The answer tells me what fraction of the original cake Sam ate. (2/3) × (12/20) = (2/3) × (3/5) = 2/5.
Sam ate 2/5 of the original cake.

9.

a. $\frac{\cancel{4}^1}{5} \times \frac{\cancel{3}^1}{\cancel{4}_1} \times \frac{2}{\cancel{3}_1} = \frac{2}{5}$	b. $\frac{11}{8} \times \frac{\cancel{6}^{\cancel{3}\,1}}{\cancel{8}_{\cancel{4}\,2}} \times \frac{\cancel{2}^1}{\cancel{3}_1} = \frac{11}{16}$
c. $\frac{9}{\cancel{10}_2} \times \frac{\cancel{5}^1}{\cancel{2}_1} \times \frac{\cancel{2}^1}{7} = \frac{9}{14}$	d. $\frac{\cancel{3}^1}{\cancel{5}_1} \times \frac{\cancel{6}^1}{\cancel{12}_2} \times \frac{\cancel{5}^1}{\cancel{3}_1} = \frac{1}{2}$
e. $\frac{\cancel{4}^1}{\cancel{5}_1} \times \frac{\cancel{9}^3}{8} \times \frac{\cancel{10}^{\cancel{2}\,1}}{\cancel{24}_{\cancel{6}\,\cancel{3}\,1}} = \frac{3}{8}$	f. $\frac{\cancel{7}^1}{\cancel{12}_2} \times \frac{3}{5} \times \frac{\cancel{6}^1}{\cancel{7}_1} = \frac{3}{10}$

10. a. Models may vary. For example: groceries / taxes

b. $490. We can see from the bar model that he uses 1/5 of his original salary for groceries (After paying 1/5 in taxes, 4/5 are left, which is four blocks in the model. Now, 1/4 of those four blocks is one block - and that is 1/5 of the original salary.) So, to find the amount he used for groceries, we divide $2,450 by 5 and get $490.

Multiplying Mixed Numbers, p. 127

1.

a. $2\frac{1}{4} \times 1\frac{1}{2}$ ↓ ↓ $\frac{9}{4} \times \frac{3}{2} = \frac{27}{8} = 3\frac{3}{8}$	b. $10\frac{1}{3} \times 2\frac{1}{2}$ ↓ ↓ $\frac{31}{3} \times \frac{5}{2} = \frac{155}{6} = 25\frac{5}{6}$
c. $5\frac{1}{5} \times \frac{1}{6}$ ↓ ↓ $\frac{26}{5} \times \frac{1}{6} = \frac{26}{30} = \frac{13}{15}$	d. $4\frac{1}{2} \times 3\frac{1}{5}$ ↓ ↓ $\frac{9}{2} \times \frac{16}{5} = \frac{144}{10} = 14\frac{2}{5}$
e. $3\frac{5}{6} \times 3\frac{1}{3}$ ↓ ↓ $\frac{23}{\cancel{6}_{3}} \times \frac{\cancel{10}^{5}}{3} = \frac{115}{9} = 12\frac{7}{9}$	f. $3\frac{1}{3} \times 5\frac{1}{10}$ ↓ ↓ $\frac{10}{3} \times \frac{51}{10} = \frac{\cancel{10}^{1}}{\cancel{3}_{1}} \times \frac{\cancel{51}^{17}}{\cancel{10}_{1}} = 17$

2. a. 3 1/30 b. 7 7/10 c. 2 3/4 d. 12 1/2

3. a. We need to find its area. The area is (5 ½ ft) × (7 ½ ft) = (11/2 ft) × (15/2 ft) = 165/4 ft^2 = 41 1/4 ft^2.
   b. The area of the room is 12 ft × 20 ft = 240 ft^2. The carpet is about 40 ft^2, so it covers about 1/6 of the floor.

4.

Cheeseball	
3	2 packages cream cheese
3 3/4	2 ½ cups shredded Cheddar cheese
2 1/4	1 ½ cups chopped pecans
1 1/2	1 teaspoon grated onion

5. Yes, it is possible.

a. Nathan's way (multiply in parts): $5 \times 4\frac{1}{7} = 5 \times 4 + 5 \times \frac{1}{7}$ $= 20 + \frac{5}{7} = 20\frac{5}{7}$	a. change the mixed number into a fraction first: $5 \times 4\frac{1}{7} = 5 \times \frac{29}{7} = \frac{145}{7} = 20\frac{5}{7}$
b. Nathan's way (multiply in parts): $8 \times 2\frac{5}{6} = 8 \times 2 + 8 \times \frac{5}{6}$ $= 16 + \frac{\cancel{8}^{4}}{1} \times \frac{5}{\cancel{6}_{3}} = 16\frac{20}{3} = 22\frac{2}{3}$	b. change the mixed number into a fraction first: $8 \times 2\frac{5}{6} = 8 \times \frac{17}{6} = 4 \times \frac{17}{3} = \frac{68}{3} = 22\frac{2}{3}$

Multiply Mixed Numbers, cont.

6. 1 ½ × 1 ½ = (3/2) × (3/2) = 9/4 = 2 1/4

7. a. 14 2/3 b. 11 7/27 c. 18 d. 12 8/9 e. 9 5/8 f. 5 1/2

8. a. The area is 8 ½ in. × 11 in. = 88 in^2 + 5 ½ in^2 = 93 ½ in^2.

   b. Now the dimensions of the paper are 7 ½ in. by 10 in. The area then becomes 7 ½ in. × 10 in. = 70 in^2 + 5 in^2 = 75 in^2.

9. a. (3 ¼ in.) × (3 ¼ in.) = (13/4 in.) × (13/4 in.) = 169/16 in^2 = 10 9/16 in^2.
   b. The outer square is 36 square inches. One-third of it would be 12 square inches. So, the smaller square occupies *less* than 1/3 of the outer square.

Multiplication as Scaling/Resizing, p. 132

1.

a. one-fourth of 24	b. one-tenth of 110 kg	c. one-fifth of one-half
$\frac{1}{4} \times \underline{24} = 6$	$\frac{1}{10} \times 110\text{ kg} = 11\text{ kg}$	$\frac{1}{5} \times \frac{1}{2} = \frac{1}{10}$
three-fourths of 24	nine-tenths of 110 kg	three-fifths of one-half
$\frac{3}{4} \times \underline{24} = 18$	$\frac{9}{10} \times 110\text{ kg} = 99\text{ kg}$	$\frac{3}{5} \times \frac{1}{2} = \frac{3}{10}$

2.

a. one-fourth of 28	b. one-sixth of 12 mi	c. one-fifth of \$400
$\frac{1}{4} \times \underline{28} = 7$	$\frac{1}{6} \times 12\text{ mi} = 2\text{ mi}$	$\frac{1}{5} \times \$400 = \$80$
one and one-fourth times 28	two and one-sixth times 12 mi	two-fifths of \$400
$1\frac{1}{4} \times 28 = 35$	$2\frac{1}{6} \times 12\text{ mi} = 24\text{ mi} + 2\text{ mi} = 26\text{ mi}$	$\frac{2}{5} \times \$400 = \$160$
two and one-fourth times 28	six and one-sixth times 12 mi	four and two-fifths times \$400
$2\frac{1}{4} \times 28 = 63$	$6\frac{1}{6} \times 12\text{ mi} = 72\text{ mi} + 2\text{ mi} = 74\text{ mi}$	$4\frac{2}{5} \times \$400 = \$1,600 + 160 = \$1760$

Multiplication as Scaling/Resizing, cont.

3.

a. ½ × ______ = ____ ½ × 50 px = 25 px 1 ½ × ______ = ______ 1 ½ × 50 px = 75 px	b. ¼ × ______ = __ ¼ × 40 px = 10 px 2 ¼ × ______ = ______ 2 ¼ × 40 px = 90 px	c. ¾ × ______ = ____ ¾ × 48 px = 36 px 1 ¾ × ______ = ______ 1 ¾ × 48 px = 84 px

d. $\frac{5}{8}$ × ______ = ____ $\frac{5}{8}$ × 40 px = 25 px 2 $\frac{5}{8}$ × ______ = ______ 2 $\frac{5}{8}$ × 40 px = 105 px	e. $\frac{3}{5}$ × ______ = ____ $\frac{3}{5}$ × 50 px = 30 px 3 $\frac{3}{5}$ × ______ = ______ 3 $\frac{3}{5}$ × 50 px = 180 px

4. a. shorter b. longer c. longer d. shorter e. longer f. shorter

5. a. 3/4 × \$6 = \$4.50
   b. 1 ½ × 8 ½ = (3/2) × (17/2) = 51/4 = 12 3/4 = \$12.75 for the nuts.
   c. 2.6 × \$11.30 = \$29.38
   d. You stay 12/30 of a month, which is 2/5 of a month. The rent is (2/5) × \$350 = \$140.

6. The new width is (2/3) × 2,400 = 1,600 pixels. The new height is (2/3) × 1,600 = 1,067 pixels.

7. The new width is 2 ¼ × 360 = (2 × 360) + (¼ × 360) = 720 + 90 = 810 pixels.
   The new height is 2 ¼ × 600 = (2 × 600) + (¼ × 600) = 1,200 + 150 = 1,350 pixels.

8.

a. Multiply the given fraction by $\frac{4}{4}$. $\frac{4}{4} \times \frac{2}{3} = \frac{8}{12}$	b. Multiply the given fraction by $\frac{3}{3}$. $\frac{3}{3} \times \frac{5}{9} = \frac{15}{27}$
c. Multiply the given fraction by $\frac{10}{10}$. $\frac{10}{10} \times \frac{2}{7} = \frac{20}{70}$	d. Multiply the given fraction by $\frac{7}{7}$. $\frac{7}{7} \times \frac{11}{12} = \frac{77}{84}$

9. a. $<$ b. $>$ c. = d. = e. $<$ f. $>$ g. = h. $>$ i. $<$

Puzzle corner. a. b. d. e. are equivalent to 5/8.

Fractions Are Divisions, p. 136

1.

a. Divide these 3 pies equally among four people. Each will get $\frac{3}{4}$ of a pie. 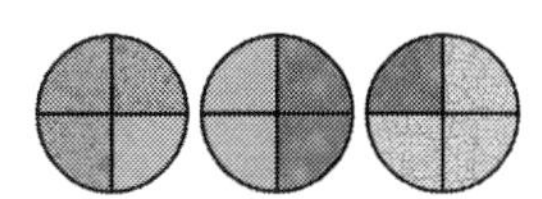	b. Divide these 2 pies equally among three people. Each will get $\frac{2}{3}$ of a pie. 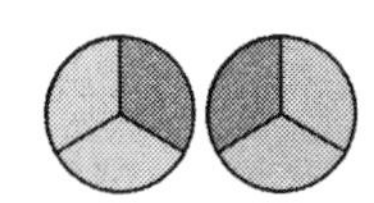
c. Divide these 5 pies equally among six people. Each will get $\frac{5}{6}$ of a pie. 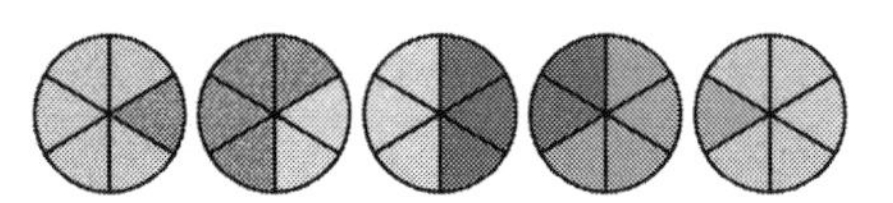	d. Divide these 5 pies equally among eight people. Each will get $\frac{5}{8}$ of a pie. 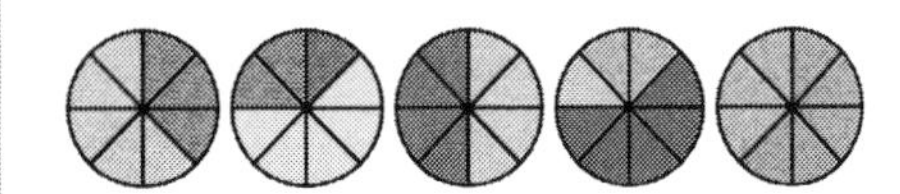

e. $\frac{3}{5}$ is the same as the division problem <u>3</u> ÷ <u>5</u>.

If three pizzas are divided equally among five people, each person gets $\frac{3}{5}$ of one pizza.

f. The answer to the division 7 ÷ 11 is $\frac{7}{11}$.	g. $8 \div 21 = \frac{8}{21}$	h. $21 \div 100 = \frac{21}{100}$

2.

a. $1 \div 5 = \frac{1}{5}$ Check: $\frac{1}{5} \times 5 = \frac{5}{5} = 1$	b. $7 \div 12 = \frac{7}{12}$ Check: $\frac{7}{12} \times 12 = \frac{84}{12} = 7$
c. $4 \div 7 = \frac{4}{7}$ Check: $\frac{4}{7} \times 7 = \frac{28}{7} = 4$	d. $7 \div 8 = \frac{7}{8}$ Check: $\frac{7}{8} \times 8 = \frac{56}{8} = 7$

3.

a. Divide 5 pies equally among four people. $5 \div 4 = \frac{5}{4} = 1\frac{1}{4}$ Each person will get <u>1 1/4</u> pies. 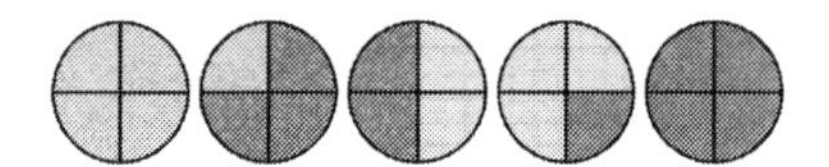	b. Five people share 6 pies equally. <u>6</u> ÷ <u>5</u> = $\frac{6}{5} = 1\frac{1}{5}$ Each person will get <u>1 1/5</u> pies. 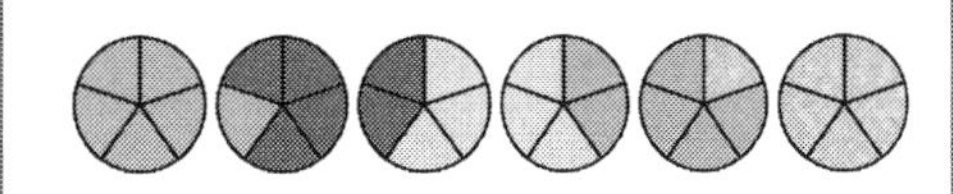

4.

a. Divide 11 pies equally among eight people. $\underline{11} \div \underline{8} = \frac{11}{8} = 1\frac{3}{8}$

Each person will get 1 3/8 pies.

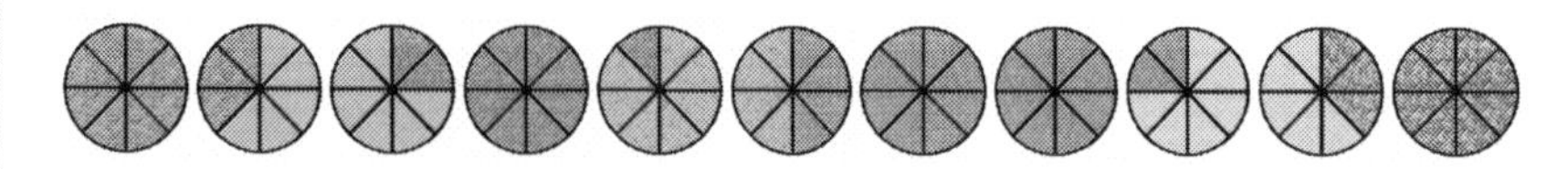

b. $\frac{17}{6}$ is the same as the division problem $\underline{17} \div \underline{6}$.

If 17 pizzas are divided equally among six people, each person gets 2 5/6 pizzas.

c. $21 \div 8 = \frac{21}{8} = 2\frac{5}{8}$

d. $46 \div 5 = \frac{46}{5} = 9\frac{1}{5}$

5.

a. $25 \div 8 = 3$ R1 $\frac{25}{8} = 3\frac{1}{8}$	b. $44 \div 5 = 8$ R4 $\frac{44}{5} = 8\frac{4}{5}$	c. $23 \div 2 = 11$ R1 $\frac{23}{2} = 11\frac{1}{2}$
d. $28 \div 3 = 9$ R1 $\frac{28}{3} = 9\frac{1}{3}$	e. $65 \div 10 = 6$ R5 $\frac{65}{10} = 6\frac{5}{10}$	f. $53 \div 9 = 5$ R8 $\frac{53}{9} = 5\frac{8}{9}$

6. a. Between 7 and 8 (it is 7 3/6).
   b. Each person gets 5/3 = 1 2/3 chocolate bars.

7. She puts 15 lb ÷ 4 = 15/4 lb = 3 3/4 lb into one bag.

8. Since 75 ÷ 4 = 18 R3, they will get 18 groups of 4, and one group of 3.

9. a. 15 kg ÷ 12 = 15/12 kg = 1 3/12 kg = 1 1/4 kg
   b. 7 in. ÷ 4 = 7/4 in. = 1 3/4 in.

10. 50 lb ÷ 9 = 50/9 lb = 5 5/9 lb. The answer is between 5 and 6.

11. 102 ÷ 11 = 9 R3. You will need 10 minibuses.

12. a. 5 L ÷ 20 = 5/20 L = 1/4 L.
    b. 1/4 L is 250 ml.

Dividing Fractions 1: Sharing Divisions, p. 140

1. (The colors may not show correctly if this is printed in black and white.)

a. $\frac{4}{6} \div 4 = \frac{1}{6}$	b. $\frac{3}{5} \div 3 = \frac{1}{5}$
c. $\frac{6}{9} \div 2 = \frac{3}{9}$	d. $\frac{6}{10} \div 3 = \frac{2}{10}$
e. $\frac{6}{12} \div 3 = \frac{2}{12}$	f. $\frac{15}{20} \div 5 = \frac{3}{20}$

2. a. (6/9) ÷ 3 = 2/9. Each person gets 2/9 of the pizza.   b. (12/20) ÷ 4 = 3/20. Each person gets 3/20 of the original cake.

3.

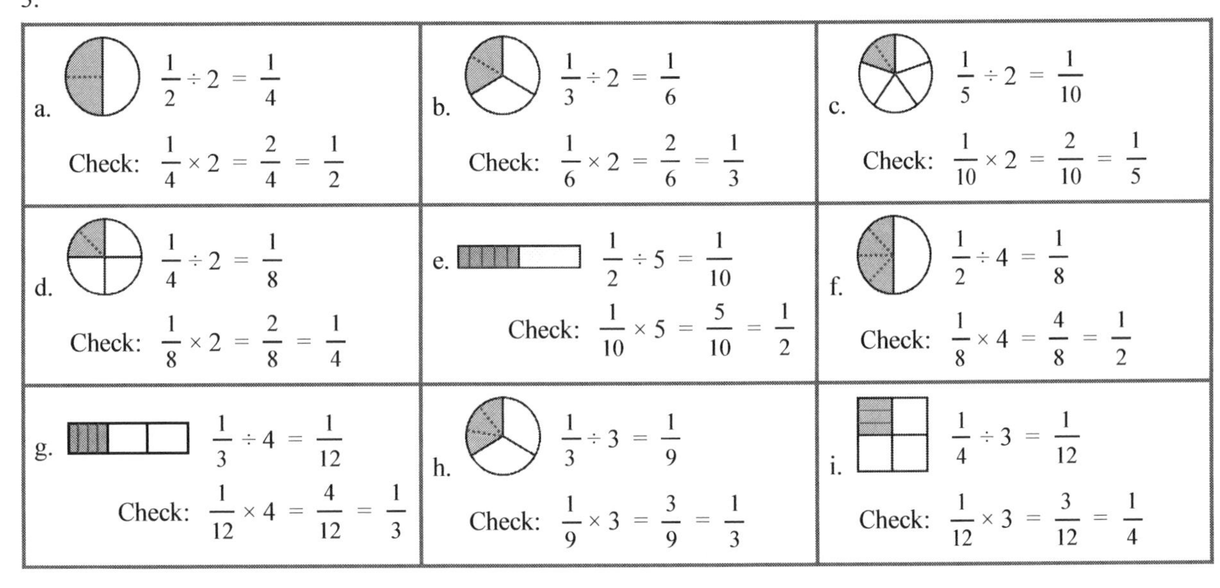

a. $\frac{1}{2} \div 2 = \frac{1}{4}$ Check: $\frac{1}{4} \times 2 = \frac{2}{4} = \frac{1}{2}$	b. $\frac{1}{3} \div 2 = \frac{1}{6}$ Check: $\frac{1}{6} \times 2 = \frac{2}{6} = \frac{1}{3}$	c. $\frac{1}{5} \div 2 = \frac{1}{10}$ Check: $\frac{1}{10} \times 2 = \frac{2}{10} = \frac{1}{5}$
d. $\frac{1}{4} \div 2 = \frac{1}{8}$ Check: $\frac{1}{8} \times 2 = \frac{2}{8} = \frac{1}{4}$	e. $\frac{1}{2} \div 5 = \frac{1}{10}$ Check: $\frac{1}{10} \times 5 = \frac{5}{10} = \frac{1}{2}$	f. $\frac{1}{2} \div 4 = \frac{1}{8}$ Check: $\frac{1}{8} \times 4 = \frac{4}{8} = \frac{1}{2}$
g. $\frac{1}{3} \div 4 = \frac{1}{12}$ Check: $\frac{1}{12} \times 4 = \frac{4}{12} = \frac{1}{3}$	h. $\frac{1}{3} \div 3 = \frac{1}{9}$ Check: $\frac{1}{9} \times 3 = \frac{3}{9} = \frac{1}{3}$	i. $\frac{1}{4} \div 3 = \frac{1}{12}$ Check: $\frac{1}{12} \times 3 = \frac{3}{12} = \frac{1}{4}$

4. a. 1/12   b. 1/20   c. 1/21   d. 1/40   e. 6/20   f. 1/28   g. 2/5   h. 1/81

5. Each child gets 1/12 lb, because (1/4 lb) ÷ 3 = 1/12 lb. In ounces, there are 4 ounces of chocolate to start with. Divide that by 3: 4 oz ÷ 3 = 4/3 oz = 1 1/3 oz. Each child gets 1 1/3 oz.

6. In liters: (1/2 L) ÷ 5 = 1/10 L. There is 100 ml of juice in each glass.

7. Add all the individual amounts, and divide by 12:
1 + 1 + 1 + 1 1/8 + 1 1/8 + 1 1/4 + 1 1/2 + 1 7/8 + 2 + 2 + 2 + 2 1/8 = 3 + 2 1/4 + 1 1/4 + 1 1/2 + 6 + 4 = 18.
Then, 18 ÷ 12 = 18/12 = 1 6/12 = 1 1/2. Each beaker would have 1 1/2 cups of oil.

8. a. 1/9   b. 1/18   c. 2/20   d. 2/11   e. 2/9   f. 2/15   g. 7/100   h. 1/22

9. a. 3/4   b. 4/5   c. 6/7   d. 9/10

10. Answers vary. Check students' answers. For example:

a. There is half of a pizza left, and three people share it equally. How much does each one get? Solution: (1/2) ÷ 3 = 1/6. Each person gets one-sixth of a pizza.

b. The pitcher is 6/8 full of juice. Mary and Mia share it equally. How much does each girl get? Solution: (6/8) ÷ 2 = 3/8. Each girl gets 3/8 of the pitcher of juice.

c. One-fourth of the fence still needs painted. Mom and Dad share the job equally. What part of the fence will each parent paint? Solution: (1/4) ÷ 2 = 1/8. Each parent will paint one-eighth of the fence.

Dividing Fractions 1: Divide a Fraction By a Whole Number, cont.

11. a. 1/16 full b. 3/16 gal c. In quarts, the full container holds 12 quarts. Being 1/16 full, it now has 12/16 qt = 3/4 qt.

12.

a. Divide 5/6 between two people. First split each piece into 2 new ones. 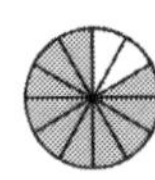 $\frac{10}{12} \div 2 = \frac{5}{12}$	b. Divide 2/3 among three people. First split each piece into 3 new ones. 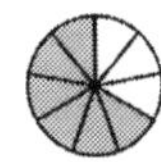 $\frac{6}{9} \div 3 = \frac{2}{9}$
c. Divide 2/3 among four people. 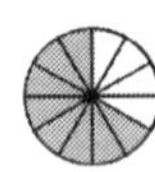 $\frac{8}{12} \div 4 = \frac{2}{12}$	d. Divide 3/4 among four people. 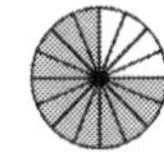 $\frac{12}{16} \div 4 = \frac{3}{16}$
e. Divide 2/5 among three people. First split each piece into 3. $\frac{6}{15} \div 3 = \frac{2}{15}$	f. Divide 4/5 among three people. 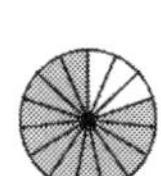 $\frac{12}{15} \div 3 = \frac{4}{15}$

Dividing Fractions 2: Fitting the Divisor, p. 145

1.

a. $2 \div \frac{1}{3} = \underline{6}$ Check: $\underline{6} \times \frac{1}{3} = \frac{6}{3} = 2$	b. $1 \div \frac{1}{4} = \underline{4}$ Check: $\underline{4} \times \frac{1}{4} = \frac{4}{4} = 1$
c. $6 \div \frac{1}{3} = \underline{18}$ Check: $\underline{18} \times \frac{1}{3} = \frac{18}{3} = 6$	d. $5 \div \frac{1}{4} = \underline{20}$ Check: $\underline{20} \times \frac{1}{4} = \frac{20}{4} = 5$
e. $5 \div \frac{1}{3} = \underline{15}$ Check: $\underline{15} \times \frac{1}{3} = \frac{15}{3} = 5$	f. $6 \div \frac{1}{2} = \underline{12}$ Check: $\underline{12} \times \frac{1}{2} = \frac{12}{2} = 6$
g. $2 \div \frac{1}{6} = \underline{12}$	h. $3 \div \frac{1}{5} = \underline{15}$

2. a. 18 b. 36 c. 32 d. 6 e. 21 f. 20 g. 6

3. a. 18 b. 20 c. 30 d. 50 e. 28 f. 32 g. 40 h. 72

4. a. 6 m ÷ (1/2 m) = 12. You can get 12 pieces.
   b. 2 c ÷ (1/4 c) = 8 servings of almonds.
   c. 5 kg ÷ (1/10 kg) = 50; He would need 50 such weights.
   d. 4 in ÷ (1/8 in) = 32 erasers

5. Answers vary. For example:

   a. How many 1/2-cup servings can you get from 2 cups of ice cream? 2 C ÷ (1/2 C) = 4. You can get 4 servings.

   b. The coach put treasure hunt questions 1/3 km apart in a 5-km track. The first question was in the very beginning, and the last one 1/3-km before the end. How many questions did he use? 5 km ÷ (1/3 km) = 15. You could of course solve this problem also by multiplying: 3 questions per km × 5 km = 15 questions.

Dividing Fractions 2: Fitting the Divisor, cont.

6. a. 6 b. 5 c. 17 d. 5 e. 8 f. 4

7.

a. How many times does (piece) go into (circles)?	
$2 \div \frac{2}{3} = 3$	$3 \times \frac{2}{3} = \frac{6}{3} = 2$

b. How many times does (piece) go into 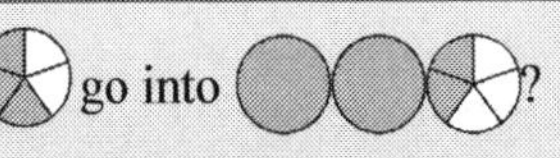?	
$2\frac{2}{5} \div \frac{3}{5} = 4$	$4 \times \frac{3}{5} = \frac{12}{5} = 2\frac{2}{5}$

c. How many times does (piece) go into 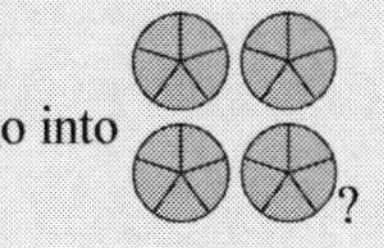?	
$4 \div \frac{2}{5} = 10$	$10 \times \frac{2}{5} = \frac{20}{5} = 4$

d. How many times does (piece) go into 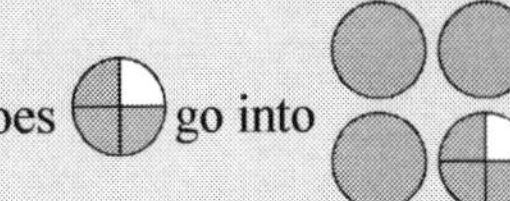?	
$3\frac{3}{4} \div \frac{3}{4} = 5$	$5 \times \frac{3}{4} = \frac{15}{4} = 3\frac{3}{4}$

8. a. She can make 6 batches. 3 C ÷ (1/2 C) = 6.
   b. She can make 5 batches. (2 1/2 C) ÷ (1/2 C) = 5.

9. She will get 18 servings. 3 × 12 ÷ 2 = 18. Or (36/12) ÷ (2/12) = 18.

10. You get 5 servings, because 1 L ÷ (2/10 L) = 5. Out of 4 liters you get four times as many servings, or 20 servings.

11. There are 10 such stretches: (2 1/2 mi) ÷ (1/4 mi) = 10.

12. a. 24 in. ÷ (1/8 in.) = 24 × 8 = 192 beads. b. Half as many, or 96.
    c. She will need 64 of each kind of bead. That pattern is 1/8 in. + 1/4 in. + 1/8 in. + 1/4 in. = 3/4 in. long. Two such patterns are 1 1/2 in. long and four such patterns are 3 inches long. We take that 8 times, and we get that 32 such patterns are 24 inches long. For 32 such patterns she will need 64 of each kind of bead.

Introduction to Ratios, p. 149

1. a. The ratio of circles to pentagons is 2 : 7. The ratio of pentagons to all shapes is 7 : 9
   7/9 of the shapes are pentagons.

   b. The ratio of diamonds to triangles is 4 : 3. The ratio of triangles to all shapes is 3 : 7
   4/7 of the shapes are diamonds.

2. a. 4:3 b. 4:7 c. 3:7

3. a. (circles) b. 8:11 c. 8/11

4.

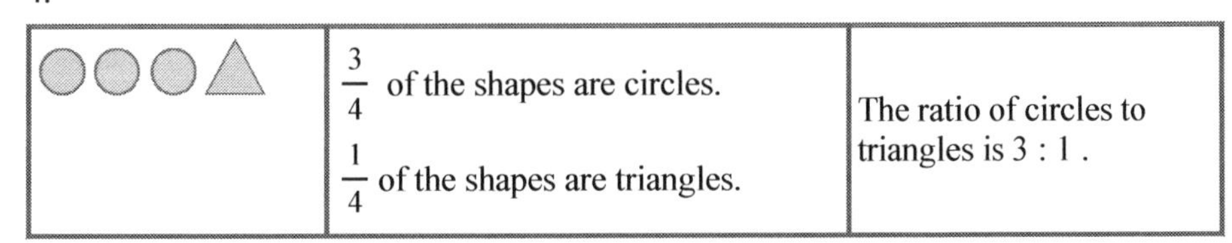

(shapes)	$\frac{3}{4}$ of the shapes are circles. $\frac{1}{4}$ of the shapes are triangles.	The ratio of circles to triangles is 3 : 1 .

5. a. 2:7 b. How many men are there? 12 How many women? 42 How many people in all? 54

Introduction to Ratios, cont.

6. a. There are 3 hearts to every 4 stars.
The ratio of hearts to stars is 12 : 16 or 3 : 4 .

b. The ratio of pentagons to circles is 8 : 12 or 2 : 3.
There are 3 circles to every 2 pentagons.

c. The ratio of diamonds to triangles is 9 : 6 or 3 : 2.
There are 3 diamonds to every 2 triangles.

7.

a. Mr. Hyde owns 1,200 acres of land. 2/3 of it is forest, and the rest is swampland. The ratio of the forest to swampland is 2 : 1.

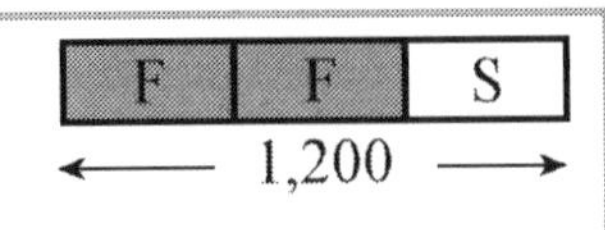

Of the land, 800 acres are forest, and 400 acres are swampland.

b. Of the 112 songs, 2/7 are slow and 5/7 are fast.

The ratio of the number of slow songs to the number of fast songs is 2 : 5.

There are 32 slow songs, and 80 fast songs.

8. a. [p|p|p|p|a] Each block represents 16 dogs. b. Sixteen are adults. c. Sixty-four are puppies.

9. Divide $147 \div 7 = 21$. Each block represents 21 marbles. There are $4 \times 21 = 84$ red marbles.

← 147 marbles →

10. a. One block represents $25. Anita gets $75.

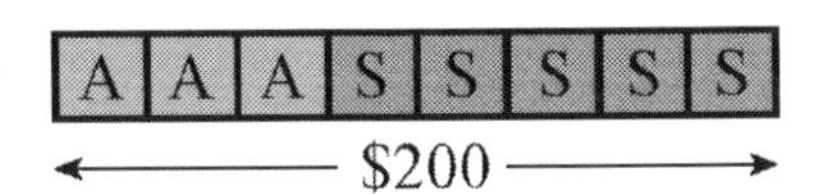

b. Shirley gets $125.

11. a. [M|M|M|M|F|F|F|F|F]
← 1,404 →

b. 4/9
c. One block represents $1{,}404 \div 9 = 156$ students. There are $4 \times 156 = 624$ male students.

12. a. [W|W|W|W|W|P]
← 102 →

b. 5:1
c. One block represents $102 \div 6 = 17$ horses. There are $5 \times 17 = 85$ white horses.

13. a. ←38→
[W|W|T|T|T|T|T]

b. 2/7 c. He has 133 marbles.

Mixed Review, p. 153

1. a. 2 3/6 b. 1 4/9 c. 2 5/12 d. 5 17/30

2. a. $0.28 = 2 \times (1/10) + 8 \times (1/100)$ b. $60.068 = 6 \times 10 + 0 \times 1 + 0 \times (1/10) + 6 \times (1/100) + 8 \times (1/1000)$

3. 400 people. People with discounted tickets: $780 \div 3 = 260$. Now we can find the number of people who paid the normal price by subtracting: $780 - 260 - 120 = 400$.

discount	seniors	normal
1/3	120	?

← 780 →

4.

Week	Weight	Weight in ounces
0	6 lb 14 oz	110
1	6 lb 12 oz	108
2	6 lb 14 oz	110
3	7 lb	112
4	7 lb 2 oz	114
5	7 lb 4 oz	116
6	7 lb 6 oz	118
7	7 lb 7 oz	119

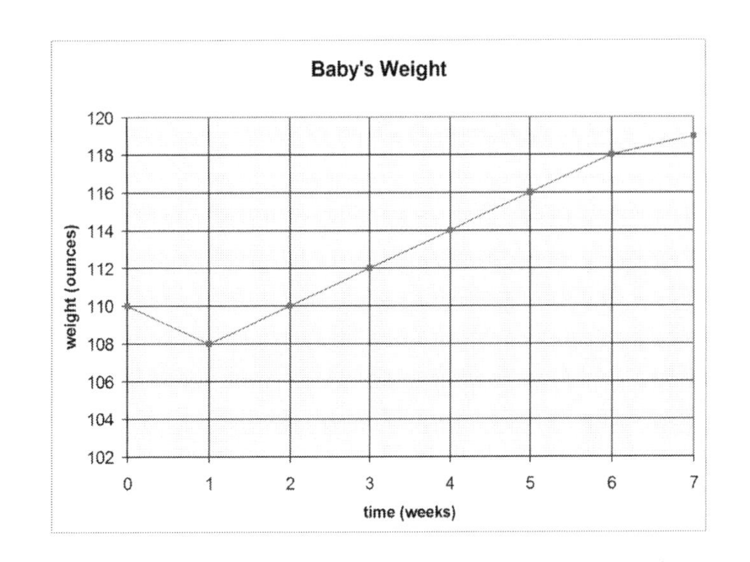

5. **The rule for *x*-values:** start at 2, and add 1 each time.
The rule for *y*-values: start at 0, and add 2 each time.

x	2	3	4	5	6	7
y	0	2	4	6	8	10

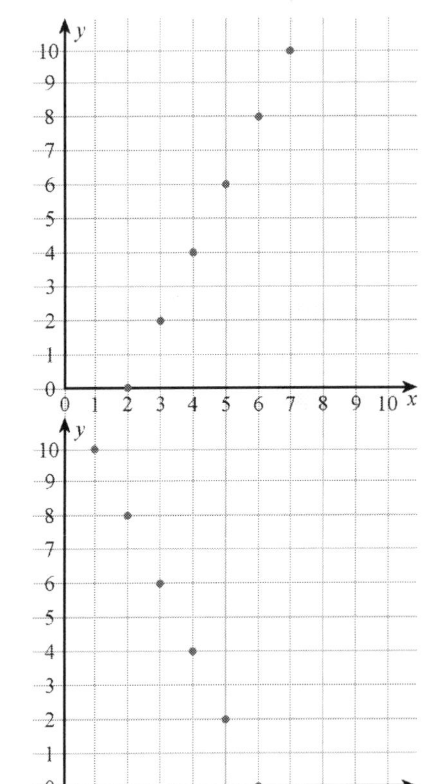

6. **The rule for *x*-values:** start at 1, and add 1 each time.
The rule for *y*-values: start at 10, and subtract 2 each time.

x	1	2	3	4	5	6
y	10	8	6	4	2	0

7. Eleven. Round the price to \$0.60. You can get 10 cans with \$6, and one more can with the \$1.

8. $4\text{ m} - 3 \times 0.82\text{ m} = 1.54\text{ m}$

9. a. $28 = 2 \times 2 \times 7$ b. $55 = 5 \times 11$ c. $84 = 2 \times 2 \times 3 \times 7$

10. a. dog b. Not possible, as the data is not numerical.

11. \$24.96. One-tenth of \$15.60 is \$1.56. Two-tenths of \$15.60 is therefore double that, or \$3.12. Subtract that from \$15.60 to find the discounted price: \$15.60 − \$3.12 = \$12.48. Jenny bought two, so her bill was \$24.96.

Mixed Review, cont.

12. a. 3,168 b. 18,216

13. a. matches with (\$170 − \$23) ÷ 7 = \$21. The answer tells how much John uses daily for groceries that are not treats.
 b. matches with \$170 − 7 × \$23 = \$9. The answer tells you how much John has to use for treats.

Puzzle corner. Solutions vary. For example:

0.4	×	0.2	= 0.08
×		×	
3	×	1.4	= 4.2
=		=	
1.2		0.28	

0.4	×	0.6	= 0.24
×		×	
0.8	×	5	= 4.0
=		=	
0.32		3.0	

Review, p. 156

1.

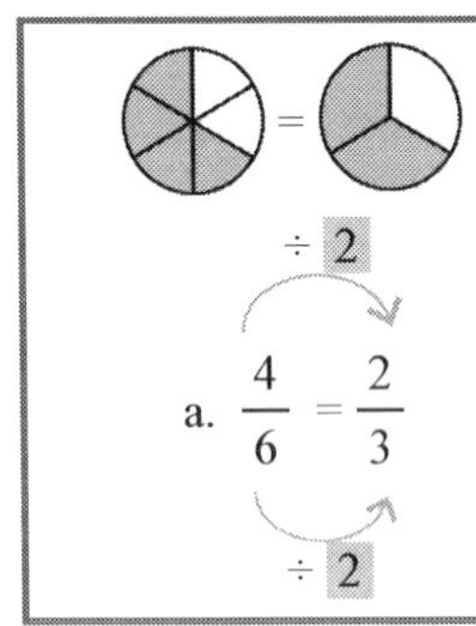

a. $\frac{4}{6} = \frac{2}{3}$ (÷ 2)

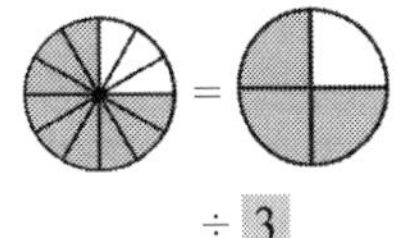

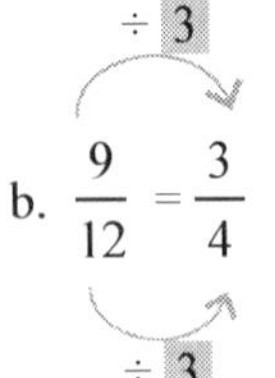

b. $\frac{9}{12} = \frac{3}{4}$ (÷ 3)

c. $\frac{24}{30} = \frac{4}{5}$

d. $3\,\frac{15}{35} = 3\,\frac{3}{7}$

e. $\frac{56}{49} = \frac{8}{7}$

f. $\frac{12}{100} = \frac{3}{25}$

g. $\frac{45}{27} = \frac{5}{3}$

h. $2\,\frac{72}{84} = 2\,\frac{12}{14} = 2\,\frac{6}{7}$

2.

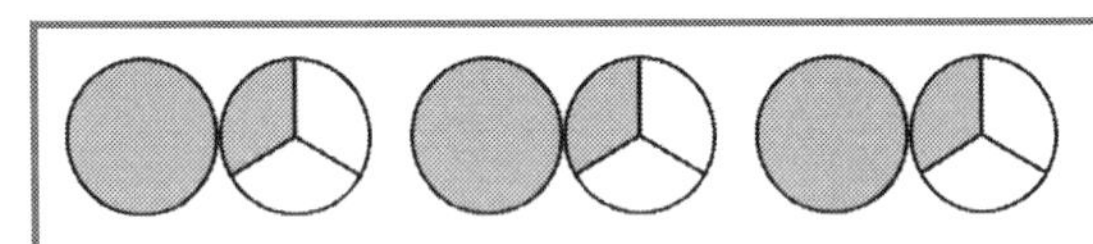

a. $3 \times 1\frac{1}{3} = 3\frac{3}{3} = 4$

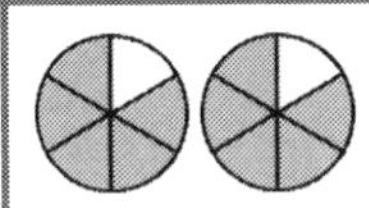

b. $2 \times \frac{5}{6} = \frac{10}{6} = 1\frac{4}{6} = 1\frac{2}{3}$

3. a. 2 4/5 b. 5/21 c. 17 1/5 d. 6 11/18

4.

a. $\frac{\overset{1}{\cancel{7}}}{\underset{2}{\cancel{14}}} \times \frac{\overset{1}{\cancel{3}}}{\underset{4}{\cancel{12}}} = \frac{1}{8}$

b. $\frac{\overset{1}{\cancel{5}}}{\underset{2}{\cancel{24}}} \times \frac{\overset{1}{\cancel{12}}}{\underset{6}{\cancel{30}}} = \frac{1}{12}$

Review, cont.

5.

a.

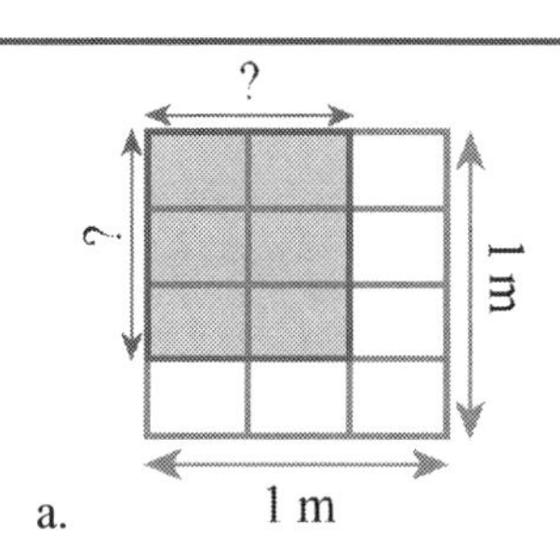

Side lengths: $\frac{2}{3}$ m and $\frac{3}{4}$ m

Area: $\frac{2}{3}\text{ m} \times \frac{3}{4}\text{ m} = \frac{6}{12}\text{ m}^2 = \frac{1}{2}\text{ m}^2$

b.

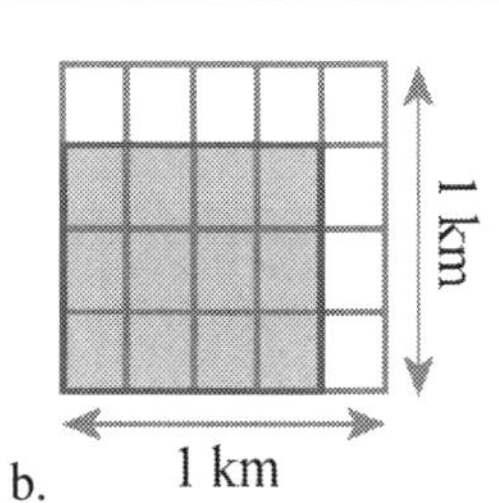

Side lengths: $\frac{4}{5}$ km and $\frac{3}{4}$ km

Area: $\frac{4}{5}\text{ km} \times \frac{3}{4}\text{ km} = \frac{12}{20}\text{ km}^2 = \frac{3}{5}\text{ km}^2$

6.

a. 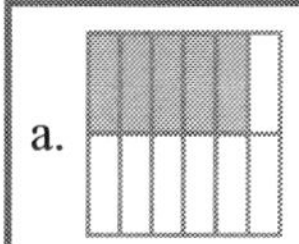 $\frac{5}{6}\text{ m} \times \frac{1}{2}\text{ m} = \frac{5}{12}\text{ m}^2$

b. 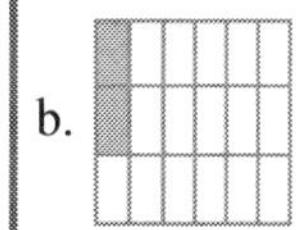 $\frac{2}{3}\text{ in.} \times \frac{1}{6}\text{ in.} = \frac{2}{18}\text{ in}^2 = \frac{1}{9}\text{ in}^2$

7. 5 × (3/4 mi) = 15/4 mi = 3 3/4 mi

8. a. Four pieces are left. Dog ate 2/3 of the 12 pieces, so 1/3 of the 12 pieces are left, which is 4 pieces.
   b. 4/20 = 1/5 of the pie is left now.

9.

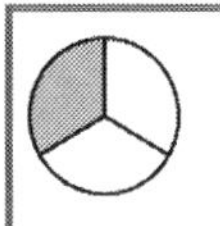

a. $1 \div 3 = \frac{1}{3}$

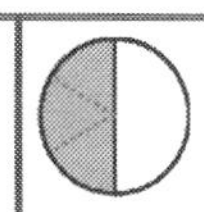

b. $\frac{1}{2} \div 3 = \frac{1}{6}$

10. a. 6 b. 16 c. 1/10 d. 1/21 e. 2 1/4 f. 1/16 g. 2/3 h. 2/10 i. 1/4

11. a. Each piece is 1 3/4 in. long: 7 in. ÷ 4 = 7/4 in. = 1 3/4 in.
    b. (4/5) × (1/3) = 4/15
    c. 11 lb ÷ 5 = 11/5 lb = 2 1/5 lb
    d. (3/4) × $24 = $18
    e. Seventy were not women. First, find 1/8 of 112, which is 14.
       Then, since 5/8 were not women, multiply that by 5 to get 70.

12. a. < b. > c. =

13. 1 − 1/3 − 1/4 = 1 − 4/12 − 3/12 = 5/12. So, 5/12 of the cake was decorated with strawberry frosting.

14. After a day, 5/6 of 30 slices were left, which is 25 slices. The family ate 1/5 of 25 slices, which is 5 slices.
    Therefore, 20 slices are now left.

15. Note: 9/16 cup or 9/16 teaspoon is not commonly found on measuring cups. You would just use a tad over 1/2 C or 1/2 teaspoon.

Brownies

2 1/4 cups sweetened carob chips
6 tablespoons olive oil
2 small eggs
3/8 cup honey
3/4 teaspoon vanilla
9/16 cup whole wheat flour
9/16 teaspoon baking powder
3/4 cup walnuts or other nuts

16. a. The 6 ½ in. by 8½ in. sheet has a greater area.
The area of the 6 ½ in. by 8½ in. sheet is (6 1/2 in.) × (8 1/2 in.) = (13/2 in.) × (17/2 in.) = 221/4 in^2 = 55 ¼ in^2.
The area of the 5¾ in. by 9 in. sheet. is (5 3/4 in.) × 9 in = (23/4 in.) × 9 = 207/4 in^2 = 51 ¾ in^2.

b. It is 3 ½ square inches larger in area. Subtract 55 ¼ in^2 − 51 ¾ in^2 = 3 ½ in^2.

Chapter 8: Geometry

Review: Angles, p. 164

1. a. 90° b. 27°

2. Check the students' answers. The acute angle should measure less than 90°, and the obtuse angle should measure more than 90°.

3. Check the students' triangles. The sum of the angle measures should be about 180°. Due to measuring errors it may also end up being 179° or 181°.

4. 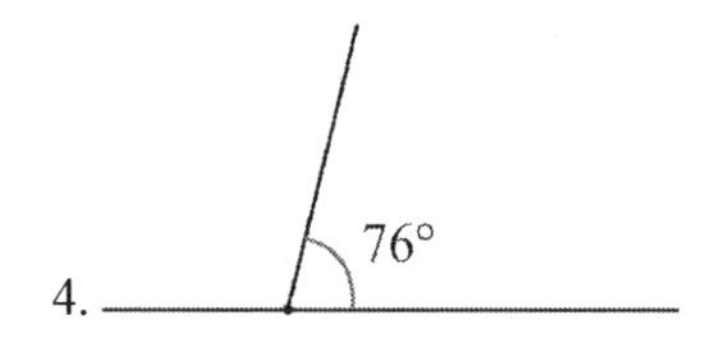

5. a. 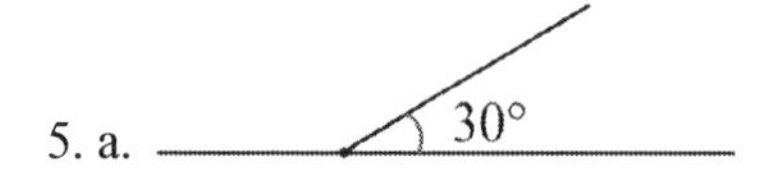

b. 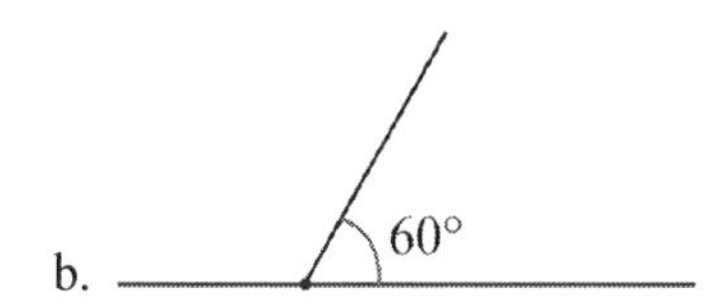

5. c. 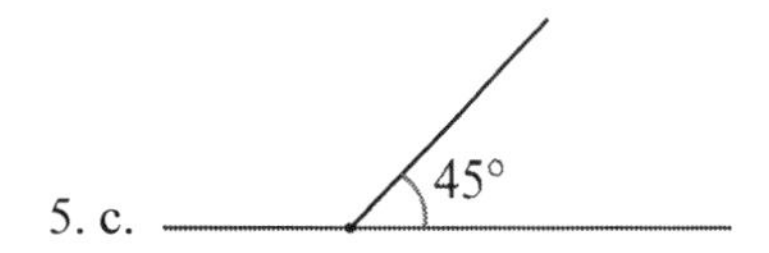

6. a. 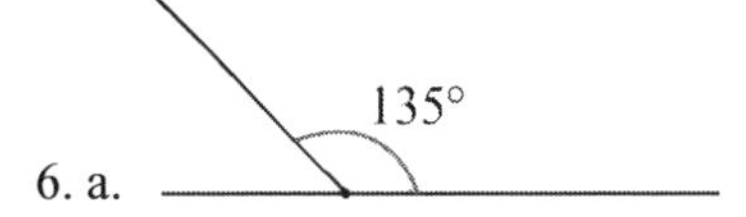

b. c. 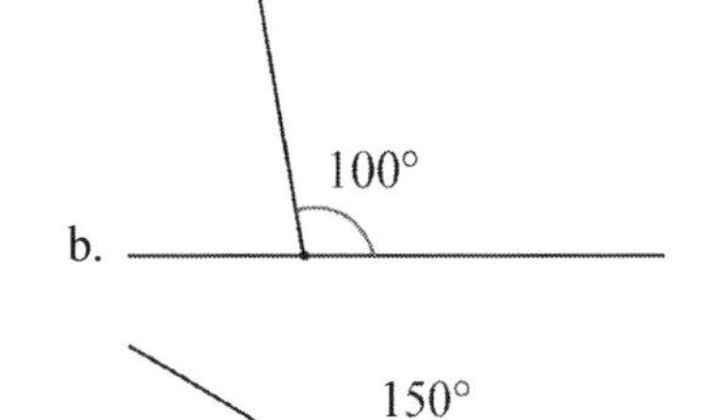

7. The estimates will vary. The exact measures are:
   a. 15° b. 94° c. 114° d. 86° e. 55° f. 121°

Review: Drawing Polygons, p. 166

1. a. A hexagon b. Answers will vary.
   c. Answers will vary.

2. 3 diagonals; 4 triangles

3. a. 4 diagonals; 5 triangles
   b. 5 diagonals; 6 triangles

4. 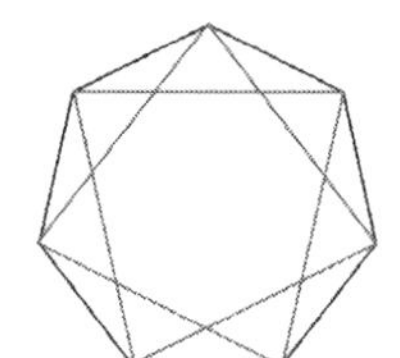

5. a. 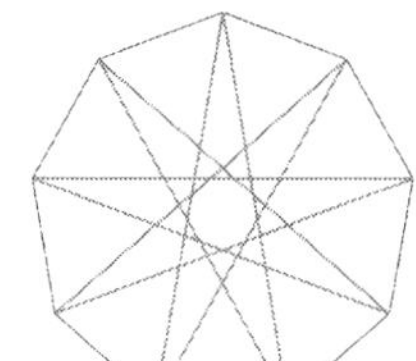

b.

c. 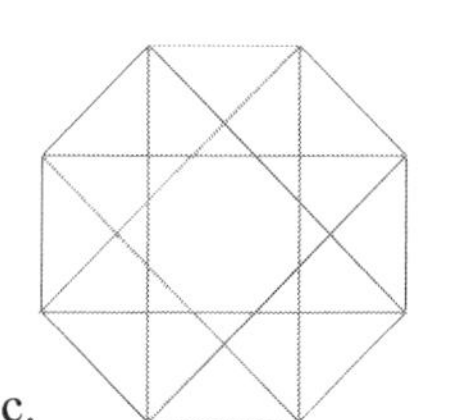

d. 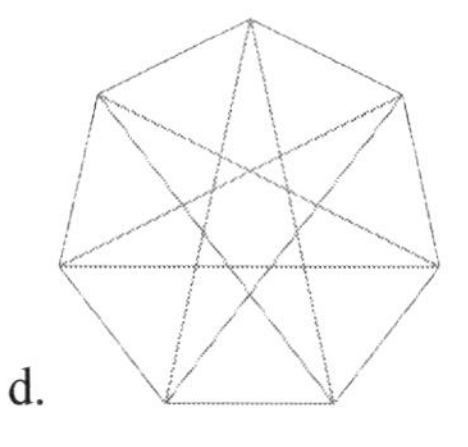

5. e. 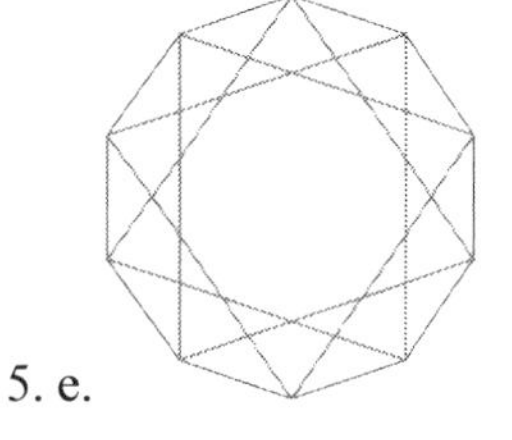

f.

6. 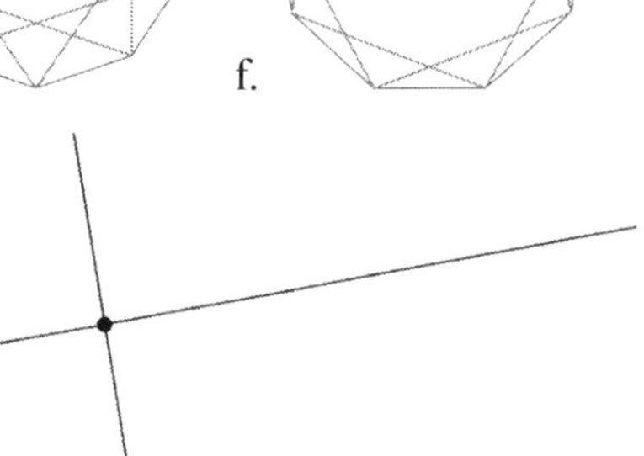

7. a. Check the student's drawing.
   b. Each diagonal measures about 4 9/16 in.
   (or 4 5/8 in. if measured to the eighth parts of an inch).
   c. All four triangles are *right* triangles.

8. a. Check that the student copied the figure accurately.
   b. Triangle A is right.
   Triangle B is obtuse.
   Triangle C is right.

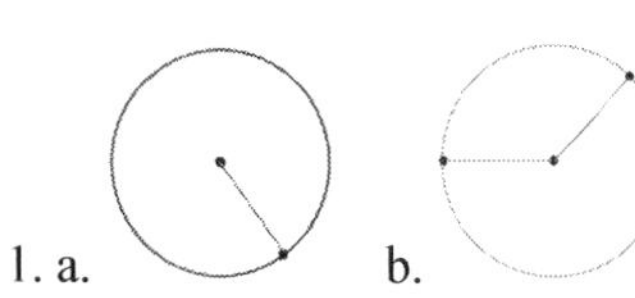

1. a. b.

c. d. e.

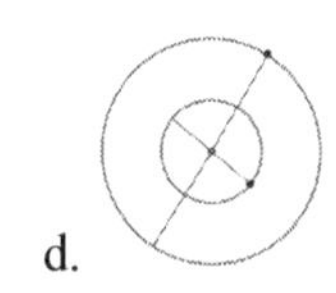

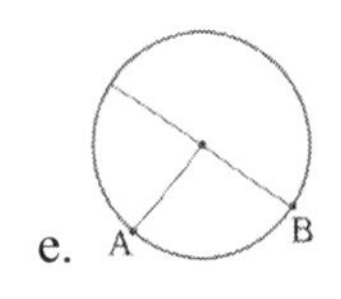

2. a. Student should practice drawing circles until the student can use the compass.
   b. c. d. Check the students' drawings.

3. a.- b. See image below.

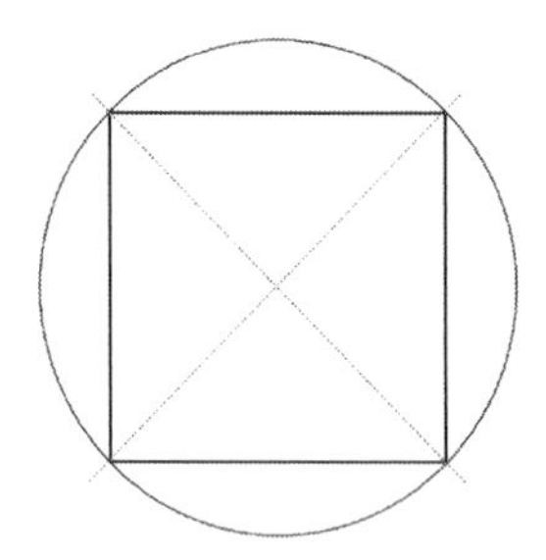

   c. The diameter of the circle

4. a. See the image below. Again, draw the two diagonals first, and use the point where they intersect as the center point of the circle.

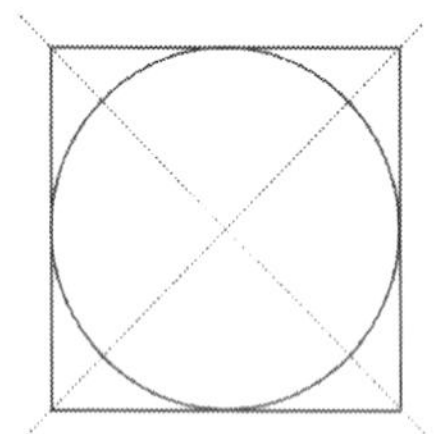

   b. The side of the square

5.

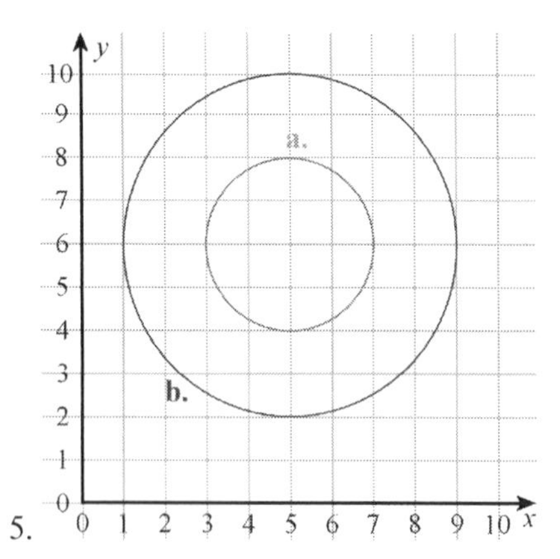

6. a. Draw a straight line, and on it, three points, equally spaced.

For example, you can choose the points to be 3 cm apart. That 3 cm will be the radius, and the points will be the center points for the circles.

After you have the points, draw the 3 circles. Use the distance between 2 points as the radius.

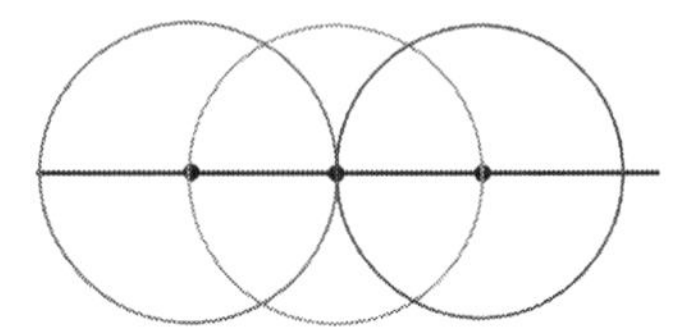

6. b. Draw a line and five dots, evenly spaced (for example, 3 cm apart).

The second and fourth of these dots are the center points of the two small circles.

Draw the two little circles using the chosen radius.

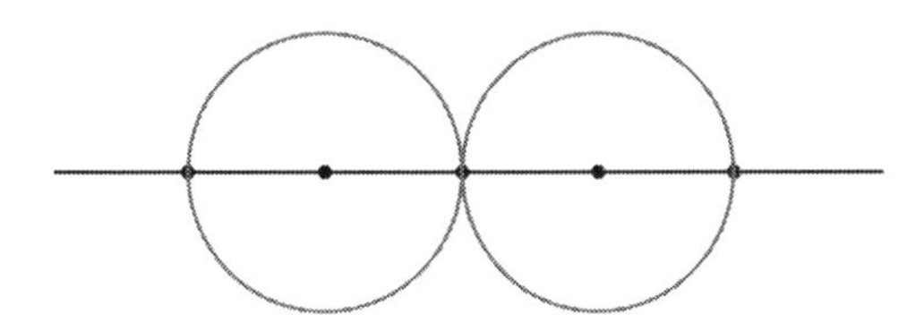

Lastly, draw the large circle using the middle dot as center point and the distance to the outer points as a radius. The radius of the small circles is exactly half of that of the big circle.

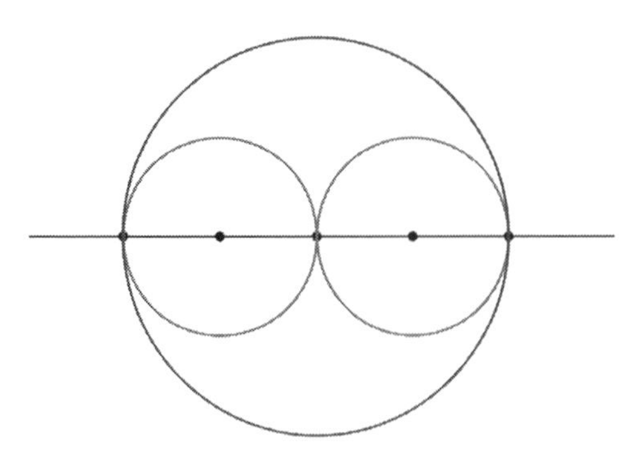

Circles, cont.

6. c. These circles are concentric: they have the same center point. The radius increases by the same amount each time. Simply draw a dot for the center point, and choose the first radius. For example, if the first circle has radius 1 cm, the next one has radius 2 cm, then 3 cm, and so on. If the first circle has radius 3/4", the next would have radius 1 1/2", the next 2 1/4", and so on.

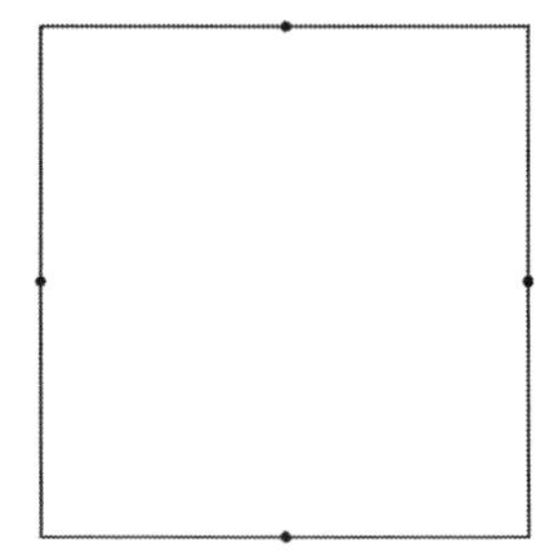

6. d. Draw first a square and mark the midpoints of its four sides (by measuring).

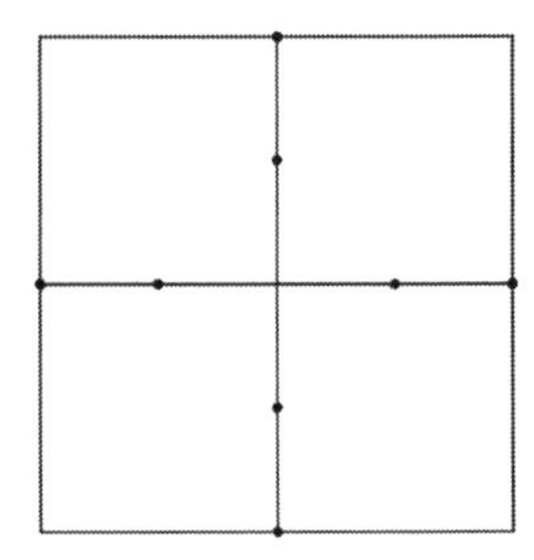

Draw the lines from midpoint to midpoint as you see in the image. The big square is now divided into quarters. Consider the smaller squares. Locate the midpoints of their sides by measuring (look at the image).

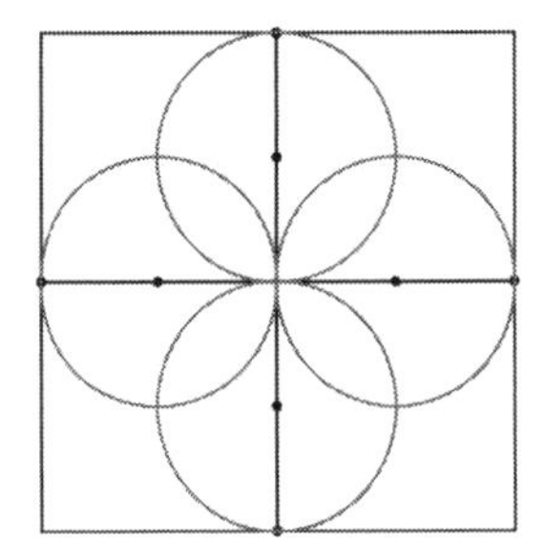

Draw the circles using those midpoints as center points for the circles.

Quadrilaterals, p. 173

1. You can draw several. You can draw a square with 2-in sides, or you can draw various rectangles where the one side is 2 inches and the other side is longer or shorter.

2. A quadrilateral with congruent sides and equal angles is called a *square*.

3. a. parallelogram b. kite
   c. trapezoid d. rhombus
   e. trapezoid f. parallelogram
   g. kite h. trapezoid
   i. rhombus j. parallelogram

4. Answers vary. For example:

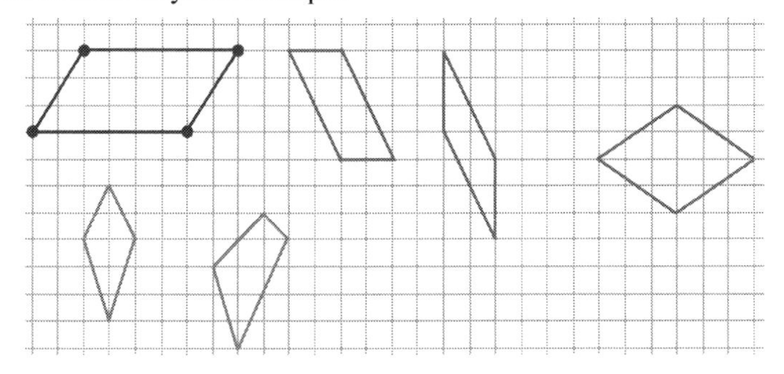

5.

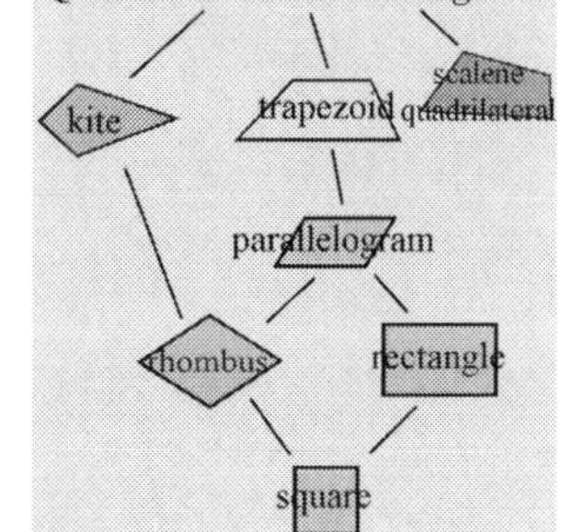

Quadrilaterals, cont.

6. a. Yes. A rhombus also has two pairs of congruent sides, like a kite. The two pairs just happen to be congruent.
   b. Yes. A square also has two pairs of congruent sides, like a kite. As in the case of the rhombus, the two pairs are congruent.
   c. No. A rectangle's adjacent sides are of different lengths—unless the rectangle is also a square.
   d. Yes. A square has a pair of parallel sides (actually two of them), so it qualifies as a trapezoid, too.
   e. No. Adjacent sides of a parallelogram are of *different* lengths, so it doesn't qualify as a kite.

7. Yes, you can, by varying how tall the trapezoid is or by varying the angles. For example (the figures are not to scale):

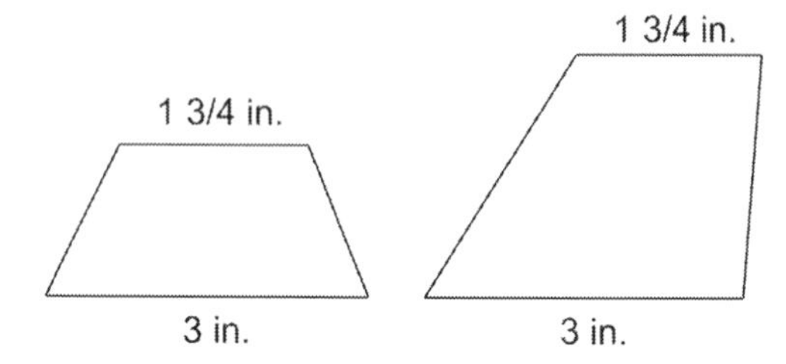

8. The adjacent sides of a parallelogram are not congruent, so it cannot be a kite nor a rhombus. However, it is also a trapezoid. (From the tree diagram in answer (5), you can see that *all* parallelograms are trapezoids).

9. It is neither a kite, nor a parallelogram, nor a rhombus.

10. She can draw several different kinds: not only a square, but also rhombi of various shapes that are not squares.

11. A square, a rectangle, a parallelogram, and a rhombus all have two pairs of congruent sides and two pairs of parallel sides.

12. Quadrilaterals puzzle: YOU GOT THE SHAPES SORTED!

Equilateral, Isosceles, and Scalene Triangles, p. 177

1. Equilateral: f, g.
   Isosceles: a, c, d (and f and g)
   Scalene: b, e, h.

2.

Triangle	Classification by the sides	Classification by the angles
a	isosceles	obtuse
d	isosceles	right
e	scalene	right
g	equilateral	acute

3.

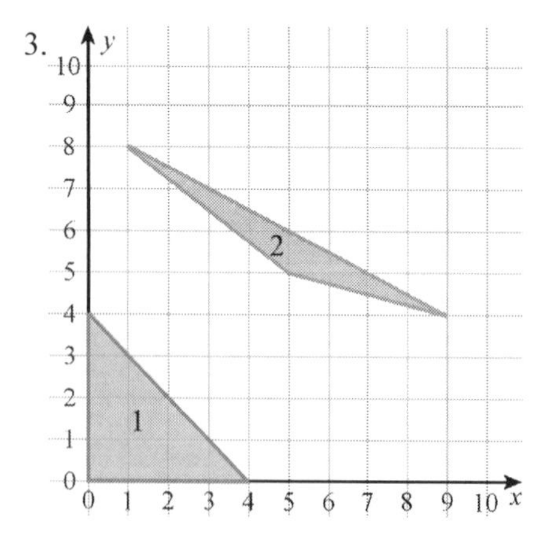

Triangle 1: right and isosceles
Triangle 2: obtuse and scalene

4. Answers vary. Check students' answers. For example:

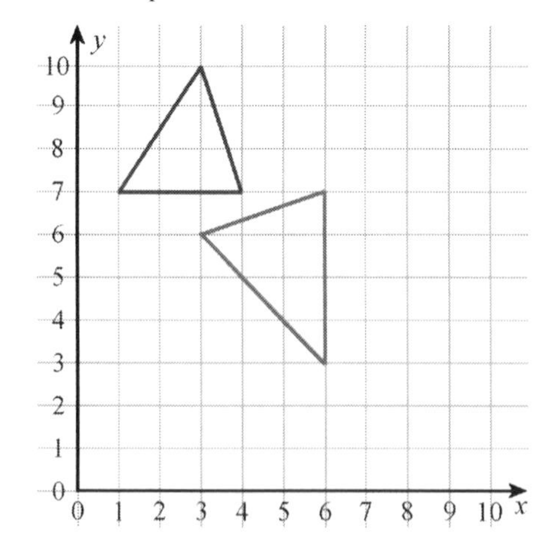

5.

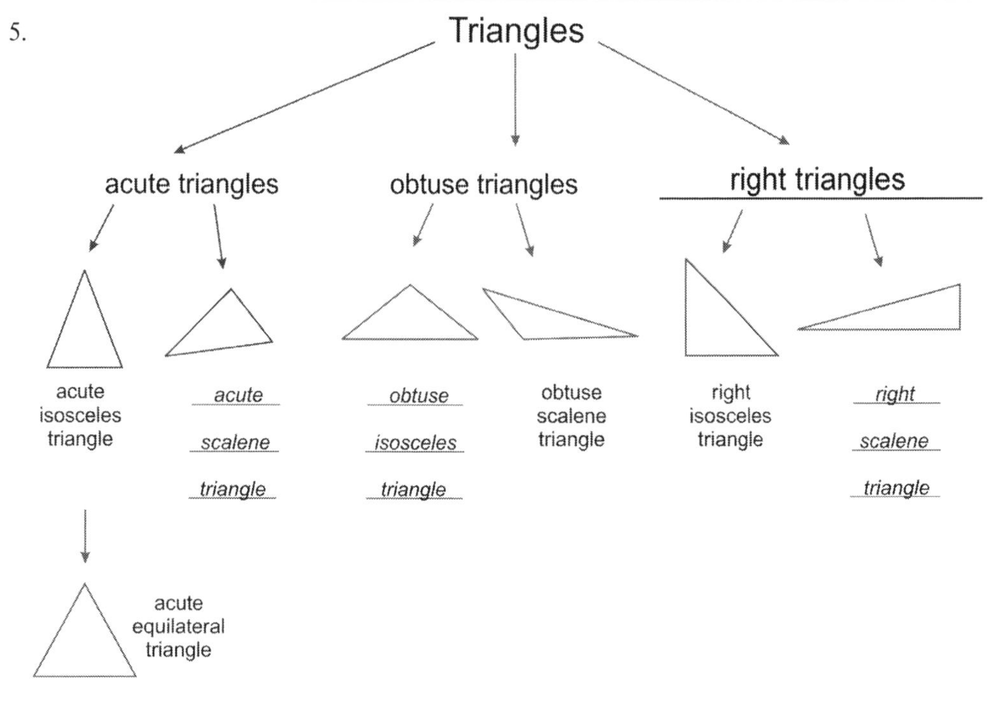

6. Answers vary. There are an innumerable number of different triangles you can draw with this information, and they are not congruent. You do not even have to have the 7-cm and 3-cm sides form the obtuse angle, though it was explained that way in the hint.

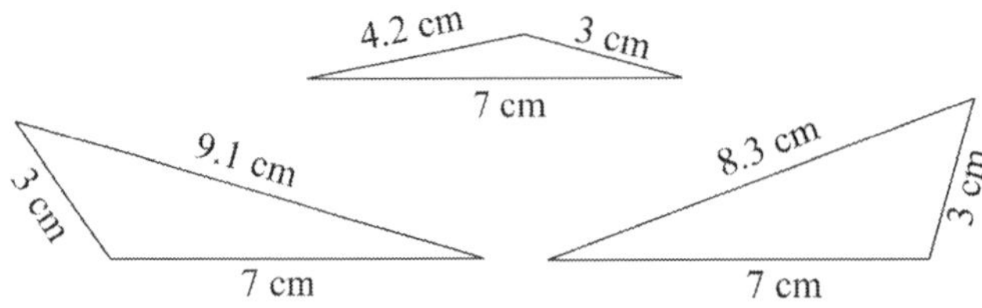

7. a. See the image on the right.

   b. The third side is 7.07 or about 7.1 cm. You can NOT draw several different-looking right triangles with two 5 cm sides. All right isosceles triangles with two 5 cm sides are identical (congruent).

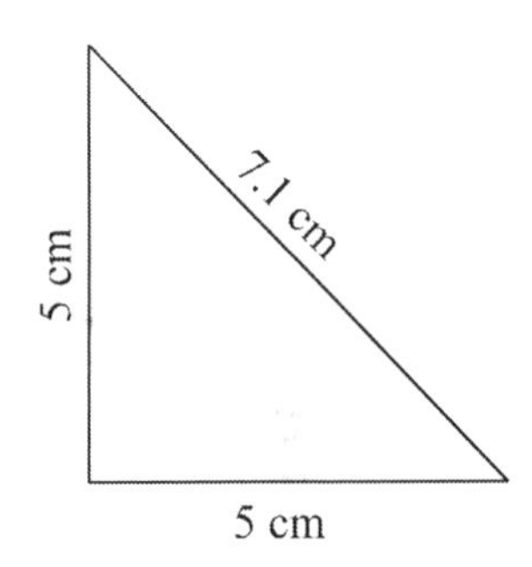

8. a. Answers will vary. Check the students' answers. The student should have drawn two lines of the same length from the same starting point with any angle between them. The third line "closes" the triangle.

   b. The angle measurements will vary. Check the students' work. Two of the angles should measure the same, and the sum should be about 180°.

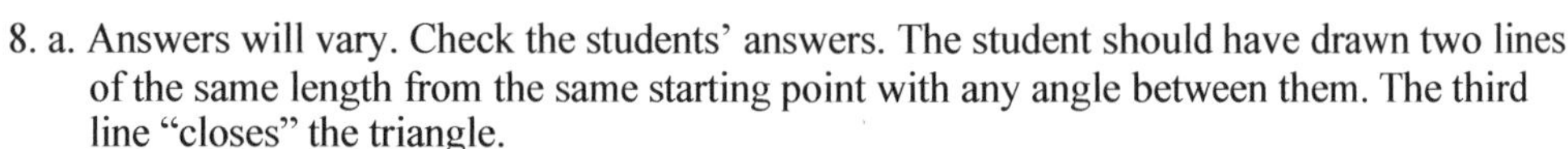

9. a. 126°, 27°, 27°. The angle sum is 180°.

   b. 80°, 50°, 50°. The angle sum is 180°.

   What do you notice? In both cases, two of the angles have the same measure.

10. The top angle is 100°. The three angle measures add up to 180°. The actual size of students' triangles may vary, but they should have the same shape as this one: →

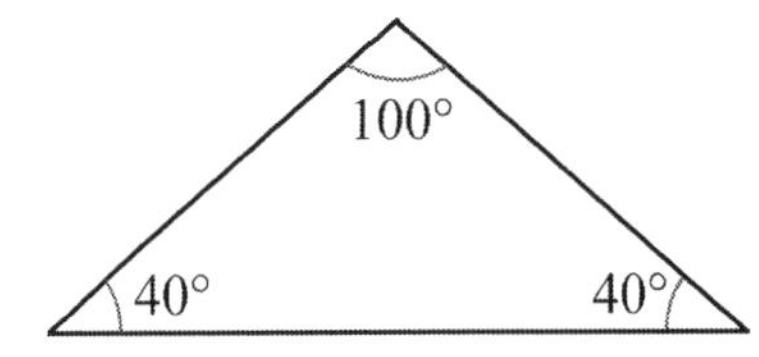

11. a. The actual size of students' triangles may vary, but they should have the same shape as this one: →

    b. The top angle is 30°. The three angle measures add up to 180°.

    c. They are not all identical. The angle measures are the same, but the side lengths can vary.

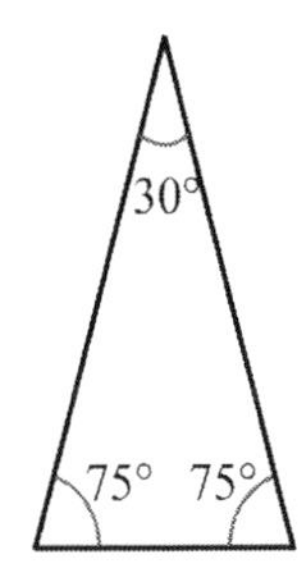

Equilateral, Isosceles, and Scalene Triangles, cont.

12. a. The actual size of students' triangles may vary, but they should have the same shape as as this one: →

50° 65° 65°

b. The base angles are 65°. The three angle measures add up to 180°.

c. They are not all identical. The angle measures are the same, but the side lengths can vary.

13. The angles measure 60°.

14. a. No, it cannot. The angles in an equilateral triangle measure 60°, so none of them can be a right angle.

b. Yes, it can. For example:

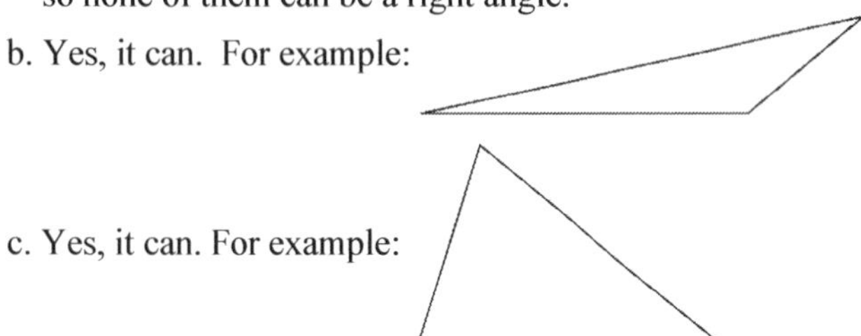

c. Yes, it can. For example:

d. Yes, it can. For example:

e. No, it cannot. The angles in an equilateral triangle measure 60°, so none of them can be obtuse.

15. Check students' drawings. Using a circle, we can find points at a certain distance from the circle's center point. In this case, we choose both circles to have the same radius, the side AB, so that we find points that are at that distance from A and from B. Where the circles intersect is a point that is at the same distance from both A and B, and that distance is the length of AB, which makes the three side lengths to be equal (congruent).

Area and Perimeter Problems, p. 182

1. a. 19 ½ sq. in. The area of the larger rectangle is 10 in. × 15 in. = 150 sq. in.
   The sides of the white rectangle are 15 − 9 ½ − 2 ¼ = 3 ¼ in. and 10 − 2 − 2 = 6 in.
   The area of the white rectangle is thus (3 ¼ in.) × 6 in = 19 ½ sq. in.
   b. The shaded area is 150 sq. in. − 19 ½ sq. in. = 130 ½ sq. in.

2. 324 cm^2. The total area of picture and frame is 38 cm × 29.5 cm = 1,121 cm^2. Subtract the width of the frame (twice, once for each side) from these dimensions to find the width and length of the picture inside the frame:
   38 cm − 2.6 cm − 2.6 cm = 32.8 cm, and 29.5 cm − 2.6 cm − 2.6 cm = 24.3 cm.
   The area of the picture inside the frame is thus 32.8 cm × 24.3 cm = 797.04 cm^2 ≈ 797 cm^2.
   The area of the frame is the difference between the total area and the area of the picture: 1,121 cm^2 − 797 cm^2 = 324 cm^2.

3. a. Area: Divide the figure into five rectangles. The middle rectangle has an area of 6 ft × 6 ft = 36 sq. ft. Each of the outer rectangles has an area of 6 ft × 4 ft = 24 sq. ft. In total, the area is then 4 × 24 sq. ft + 36 sq. ft = 132 sq. ft.
   b. Perimeter: 4 × 6 ft + 8 × 4 ft = 24 ft + 32 ft = 56 ft.

4. 10 cm. Subtract the long side twice from the perimeter: 42 cm − 11 cm − 11 cm = 20 cm. This is now twice the length of the short side. Take half of that to find the length of the short side, 10 cm.

5. The areas of the four component rectangles are: 6 × 12 = 72 cm^2; 9 × 6 = 54 cm^2; 6 × 12 = 72 cm^2; and 6 × 9 = 54 cm^2.
   So, the total area is 72 cm^2 + 54 cm^2 + 72 cm^2 + 54 cm^2 = 252 cm^2.
   The perimeter is 3 + 3 + 6 + 6 + 9 + 6 + 6 + 9 + 6 + 3 + 6 + 6 + 9 + 6 + 6 = 90 cm.

6. 125 ft. Again, subtract the given side *twice* from the perimeter, and you will have twice the other side:
   910 ft − 330 ft − 330 ft = 250 ft. Half of 250 ft is 125 ft.

7. a. (1/2 mi) ÷ 4 = 1/8 mi.
   b. To convert from miles to feet, multiply by 5,280: (1/8) × 5,280 = 5,280 / 8 = 660 ft.
   You can also solve this by converting the perimeter of 1/2 mile to feet, and dividing that by 4.

Area and Perimeter Problems, cont.

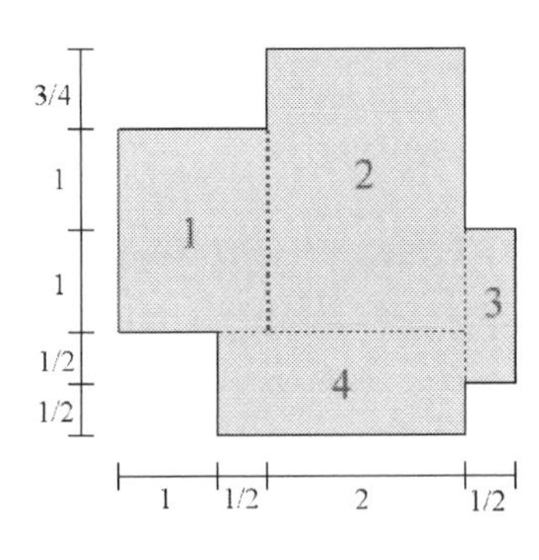

8. a. The areas of the rectangles 1, 2, 3, and 4 illustrated at right are:
   Rectangle 1: $(1 + 1) \times (1 + \frac{1}{2}) = 2$ in $\times 1\frac{1}{2}$ in $= 3$ sq. in.
   Rectangle 2: $(\frac{3}{4} + 1 + 1) \times 2 = 2\frac{3}{4}$ in $\times 2$ in $= 5\frac{1}{2}$ sq. in.
   Rectangle 3: $(1 + \frac{1}{2}) \times \frac{1}{2} = 1\frac{1}{2}$ in $\times \frac{1}{2}$ in $= \frac{3}{4}$ sq. in.
   Rectangle 4: $1 \times (\frac{1}{2} + 2) = 1$ in $\times 2\frac{1}{2}$ in $= 2\frac{1}{2}$ sq. in.

   The total area is the sum:
   3 sq. in. + $5\frac{1}{2}$ sq. in. + $\frac{3}{4}$ sq. in. + $2\frac{1}{2}$ sq. in. = $11\frac{3}{4}$ sq. in.

   b. The perimeter is:
   $1\frac{1}{2} + \frac{3}{4} + 2 + 1\frac{3}{4} + \frac{1}{2} + 1\frac{1}{2} + \frac{1}{2} + \frac{1}{2} + 2\frac{1}{2} + 1 + 1 + 2 = 15\frac{1}{2}$ in.

Volume, p. 185

1. a. V = 6 cubic units   b. V = 16 cubic units   c. V = 12 cubic units
   d. V = 32 cubic units   e. V = 36 cubic units   f. V = 28 cubic units
   g. V = 12 cubic units   h. V = 20 cubic units   i. V = 16 cubic units

2. a. V = 3 in^3 or 3 cu. in.   b. V = 30 ft^3 or 30 cu. ft.   c. V = 12 cm^3
   d. V = 6 m^3 e. V = 13 cm^3 f. 19 in^3 or cu. in.

3.

	a.	b.	c.	d.
Cubes in the bottom layer	*8*	12	21	6
Height	*4*	4	3	6
Volume	*32*	48	63	36

4. There are $10 \times 3 \times 3 = 90$ cubes that fit into the box. So, the volume of the box is 90 cubic inches.

5.

	a.	**b.**	**c.**	**d.**
Width:	3	3	3	4
Depth:	3	2	2	4
Height:	2	3	6	4
Volume:	18	18	36	64

6. a. $V = 3 \times 3 \times 1 = 9$ cubic units   b. $V = 3 \times 2 \times 2 = 12$ cubic units
   c. $V = 3 \times 2 \times 5 = 30$ cubic units   d. $V = 4 \times 4 \times 3 = 48$ cubic units

7. Check students' answers. The orientation of the prisms may vary. For example:

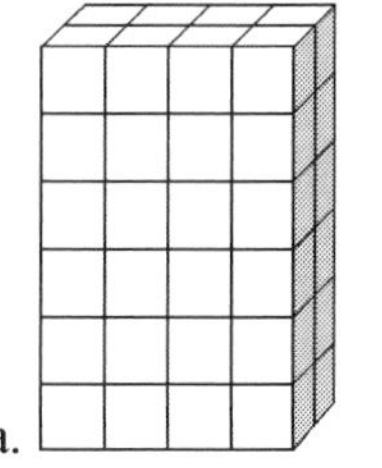
a.

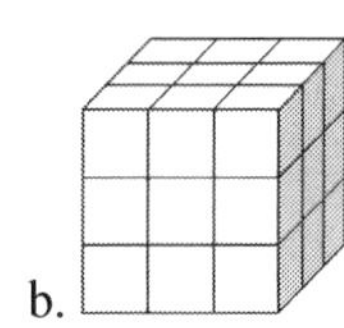
b.

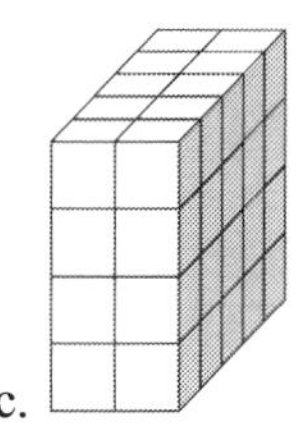
c.

8. Both are correct. The two prisms just have a different orientation. Both have the same volume, and both have dimensions of 5, 3, and 2 units.

9. c. $2 \times 2 \times 2 + 2 \times 2 \times 5$

Volume of Rectangular Prisms (Cuboids), p. 190

1. The order of factors in the multiplications may vary.

   a. V = *5 in.* × 7 in. × 2 in. = 70 in^3

   b. A = 25 ft × 20 ft = 500 ft^2

   V = 25 ft × 20 ft × 9 ft = 4,500 ft^3

2. a. 8 cm × 4 cm × 10 cm = 320 cm^3
   b. 16 in^2 × 6 in. = 96 in^3

3. V = 70 ft × 25 ft × 6 ft = 70 × 150 ft^3 = 10,500 ft^3

4. Its depth is 2 cm, because 5 cm × 3 cm × 2 cm = 30 cm^3.

5. Check students' answers. The measuring could be done in inches or in centimeters, and the volume will be correspondingly in cubic inches or cubic centimeters.

6. 2 inches

7. a. Answers vary. Check the students' work. One possibility is: Width 4 in., height 4 in., and depth 4 in. Another is: Width 2 in., height 8 in., and depth 4 in.

   b. Answers vary. Check the student's work. One possibility is: Width 12 cm, height 10 cm, and depth 10 cm. Another is: Width 50 cm, height 8 cm, and depth 3 cm.

8. Its height is 6 inches, because 5 in. × 6 in. × 3 in. = 180 in^3. Or, you can use division and solve 180 ÷ (5 × 6) = 6.

9. Total volume is 28,000 cm^3. The bottom box: V = 50 cm × 20 cm × 20 cm = 20,000 cm^3.
   Top box: V = 20 cm × 20 cm × 20 cm = 8,000 cm^3. Notice the depth of the top box is the same as the depth of the bottom box (20 cm). Total volume is found by adding and is 28,000 cm^3.

10. a.

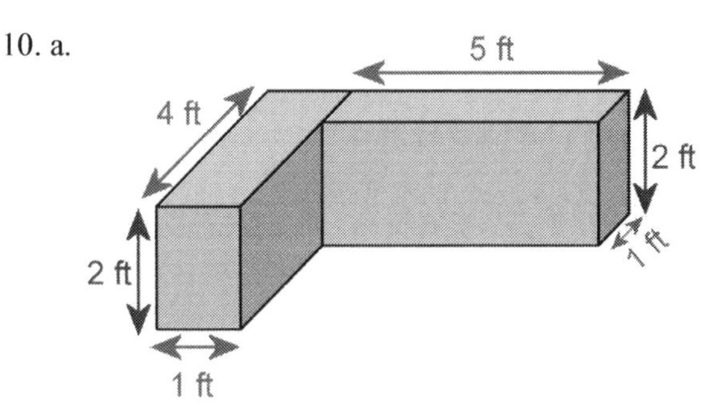

   b. The volume is 1 ft × 4 ft × 2 ft + 1 ft × 5 ft × 2 ft = 8 ft^3 + 10 ft^3 = 18 ft^3.

11. The total volume is 936 m^3. The volume of the 'bottom' part is 12 m × 8 m × 6 m = 576 m^3. The volume of the upper part is 5 m × 8 m × 9 m = 360 m^3.
    The total volume is 576 m^3 + 360 m^3 = 936 m^3.

12. a. 40 cm × 27 cm × 22 cm = 23,760 cm^3.
    b. Since it is 1/4 full, the volume of water is 23,760 cm^3 ÷ 4 = 5,940 cm^3.

13. a. 12 ft × 18 ft × 8 ft = 1,728 cu. ft.
    b. The room loses 12 ft × 18 ft × 1 ft = 216 cu. ft. of volume.

14. a. The total volume is 107,400 in^3.
    Part on the left:
    V = 38 in. × 30 in. × 50 in. = 57,000 in^3.
    Part on the right:
    V = 40 in. × 30 in. × 42 in. = 50,400 in^3.
    The total volume is 107,400 in^3.

    b. 62.153 ft^3.

Puzzler Corner. The edge length of the larger cube is 10 in. Therefore, the edge length of the smaller cube is 5 in., and its volume is 5 in. × 5 in. × 5 in. = 125 cubic inches. The total volume of the two boxes is then 1,125 cubic inches.

A Little Bit of Problem Solving, p. 194

1. a. The area is 484 m^2 (22 m × 22 m).
   b. Square number 50, which has a side 100 m long.

Square number	Side Length	Area
1	2 m	4 m^2
2	4 m	16 m^2
3	6 m	36 m^2
4	8 m	64 m^2
5	10 m	100 m^2
6	12 m	144 m^2

2. a. Student sketches may vary, but should show a small rectangle inside a large rectangle, and the dimensions marked.

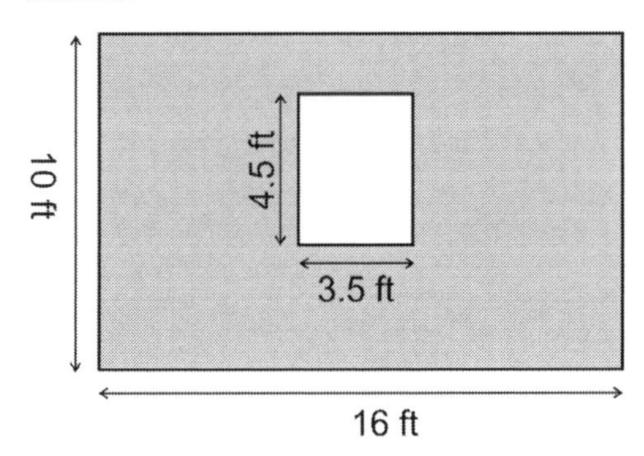

   b. The area of the window is 3.5 ft × 4.5 ft = 15.75 ft^2.

   c. The area of the actual wall is
   16 ft × 10 ft − 15.75 ft^2 = 160 ft^2 − 15.75 ft^2
   = 144.25 ft^2.

2. d. Since 1 gallon covers 350 sq. ft., it is obviously enough to cover the whole wall.

   e. 1 quart will cover 1/4 of 350ft^2, which is 87.5 ft^2. Then, 2 quarts will cover double that, or 175 ft^2. So 2 quarts is enough.

3. a. V = 4 ft × 3 ft × 3 ft = 36 cu. ft.
   b. 32 cu. ft.

4. 40/105 = 8/21

5. (5 ft × 4 ft)/(15 ft × 20 ft) = (20 sq. ft)/300 (sq. ft) = 1/15.

6. a. Check students' triangles. The triangle should look like this (image not to scale):

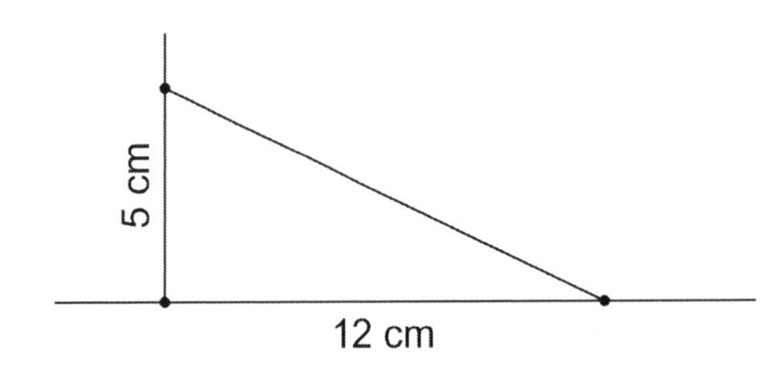

   b. 30 cm. (The third side of the triangle should measure 13 cm, or close.)

Mixed Review, p. 196

1. First, subtract \$80 − \$12 = \$68. Then take half of that: \$34. Angela got \$34 and Eric got \$46.

Angela [\$34]
Eric [\$34][\$12]
} \$80

2. \$6. One orchid costs \$8, and one daisy costs \$2. The price difference is \$6.

3. a. 215. Check: 46 × 215 = 9,890
   b. 1.1 Check: 65 × 1.1 = 71.5

4. a. 0.75 b. 0.64 c. 0.09 d. 7.2 e. 0.01 f. 4.2
   g. 0.2 h. 8 i. 0.9 j. 2 k. 200 l. 0.008

5.

a. $\frac{6}{5} = 1\frac{1}{5}$	b. $\frac{2}{21}$	c. $\frac{3}{44}$
d. $\frac{5}{9}$	e. 15	f. 15
g. $\frac{1}{10}$	h. $\frac{1}{30}$	i. $\frac{7}{5} = 1\frac{2}{5}$
j. $\frac{4}{9}$	k. $\frac{40}{3} = 13\frac{1}{3}$	l. $\frac{62}{9} = 6\frac{8}{9}$

6. Four jars: 4 × (1 3/8 in.) = 4 12/8 in. = 5 1/2 in.

7. a. $x = 4 \times 14 = 56$ b. $x = 1{,}500 \div 6 = 250$

Mixed Review, cont.

8. See the recipe on the right.

   What do you think she should do with the eggs?

   She should use 1 egg. Technically, 2/3 of 2 eggs is 1 1/3 eggs, but that is not practical to use.

> Pancakes
>
> 2 2/3 dl water
> 1 egg
> 2 dl whole wheat flour
> (pinch of salt)
> 33 g butter for frying

9.

a. $4\frac{1}{2} - 1\frac{3}{8}$

$= 4\frac{4}{8} - 1\frac{3}{8} = 3\frac{1}{8}$

b. $3\frac{1}{3} - 2\frac{7}{12}$

$= 3\frac{4}{12} - 2\frac{7}{12} = \frac{9}{12} = \frac{3}{4}$

10. a. 3:8 b. 5:8 c. 5:3

11. a.

Museum's visitors			
Day	**Adults**	**Children**	***Total Visitors***
Monday	29	14	43
Tuesday	23	10	33
Wednesday	34	18	52
Thursday	38	19	57
Friday	35	19	54
Saturday	57	25	82
Sunday	63	31	94
Totals	279	136	415

b. Sunday is the busiest day with 94 visitors, and Tuesday the least busy with 33 visitors . The difference in the total visitor count between those two days is 94 − 33 = 61.

c. 279 ÷ 7 = 39.9

d. 136 ÷ 7 = 19.4

e.

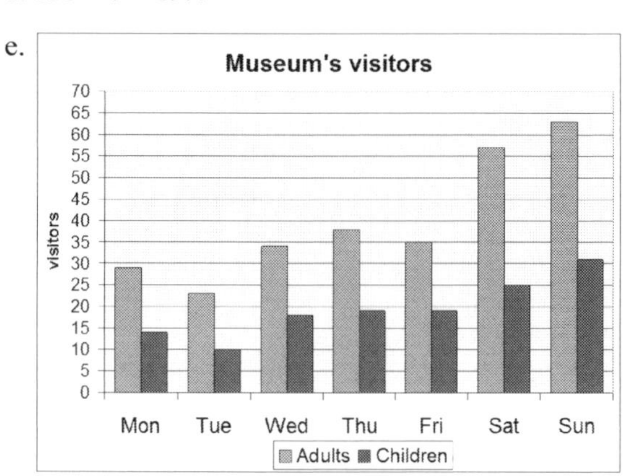

Review, p. 199

1. a. Angles 64°, 64°, and 52°. It is an acute isosceles triangle.
   b. Angles 30°, 125°, and 25°. It is an obtuse scalene triangle.

2. a. Image on the right (not to scale):
   b. The top angle is 80°.
   c. Perimeter: 70 mm + 54 mm + 54 mm = 178 mm

50° 50°
7 cm

3. Starting from the top left corner and going clockwise, the perimeter is:
   (¾ + ½ + 2) + (1 + 1 + ½ + ½ + ½ + ½) + (2 + ½ + ½ + ½ + ¾) + 1 + 1

   = 3 ¼ + 4 + 4 ¼ + 2 = 13 ½ inches.

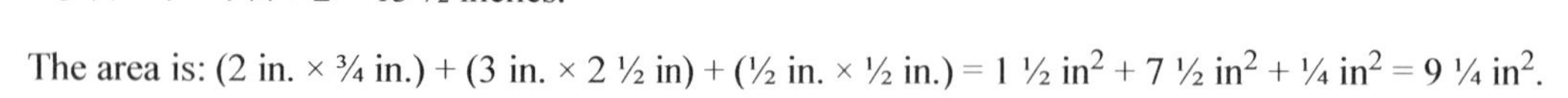

   The area is: (2 in. × ¾ in.) + (3 in. × 2 ½ in) + (½ in. × ½ in.) = 1 ½ in^2 + 7 ½ in^2 + ¼ in^2 = 9 ¼ in^2.

Review, cont.

4. a. rhombus b. rectangle c. square d. kite e. parallelogram f. scalene quadrilateral g. trapezoid

5. a. rectangle b. rhombus c. trapezoid

6. a. a pentagon b. c. d. There are several ways to draw the diagonals:

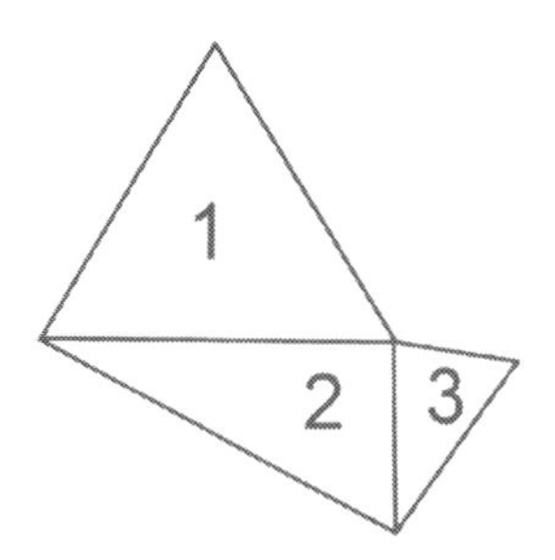

Triangle 1: Acute isosceles triangle
Triangle 2: Right scalene triangle
Triangle 3: Acute scalene triangle

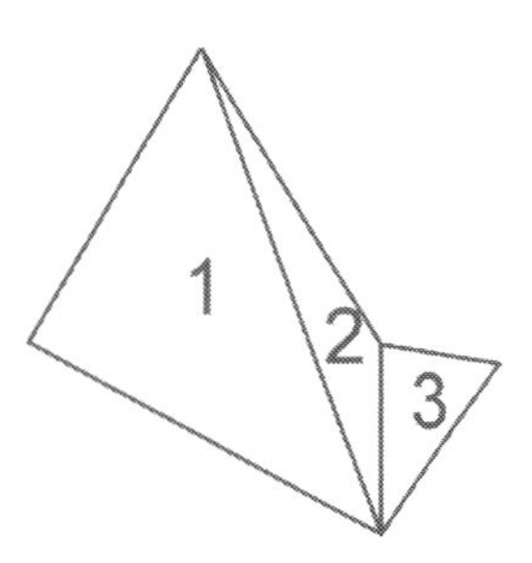

Triangle 1: Acute scalene triangle
Triangle 2: Obtuse scalene triangle
Triangle 3: Acute scalene triangle

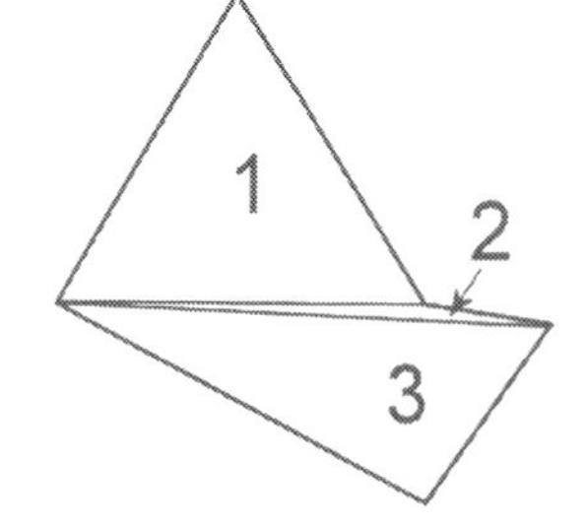

Triangle 1: Acute isosceles triangle
Triangle 2: Obtuse scalene triangle
Triangle 3: Obtuse scalene triangle

7. Answers vary. Check students' answers. For example: a.

b.

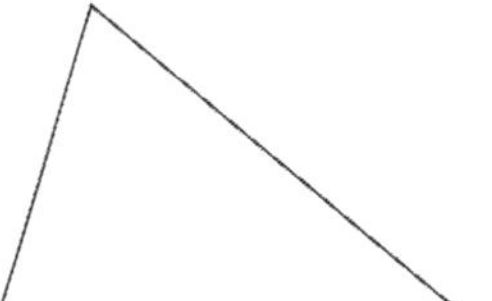

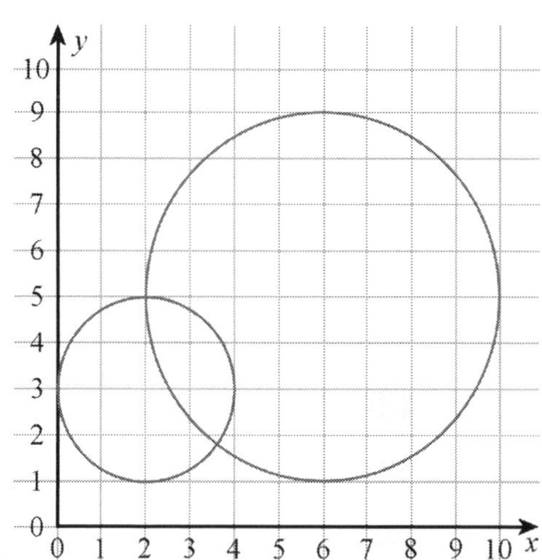

8. a. b.

9. 4 cm. The area of the bottom is 2 cm × 4 cm = 8 cm^2, so the missing dimension is 32 cm^3 ÷ 8 cm^2 = 4 cm.

10. Three boxes. One box has a volume of 6 in. × 3 in. × 2 in. = 36 cubic inches. You will need three of them to have a total volume of 108 cubic inches.

11. a. The figure has 4 × 5 × 2 = 40 cubes, so its volume would be 40 cubic inches.
    b. In this case, each little cube would have a volume of 2 in. × 2 in. × 2 in. = 8 cu. in. There are 40 cubes, so their total volume is 40 × 8 cu. in. = 320 cu. in. Or, you can calculate the volume by first calculating the three dimensions: the length is 10 inches, the depth is 4 inches, and the height is 8 inches, so the volume is then 10 in. × 4 in. × 8 in. = 320 cu. in.

Puzzle corner. The edge length of the cube must be 4 cm. Therefore, its volume is 4 cm × 4 cm × 4 cm = 64 cm^3.

Test Answer Keys

Math Mammoth Grade 5 Tests Answer Key

Please use the tests provided as you see fit. I have tried to write them so they can be used to assess the student's knowledge of the main concepts and skills in each chapter. You can, however, use them as an additional review lesson instead (just ignore the word "test" in the title).

I do not personally feel that students necessarily need to be tested often in a homeschool situation, because the parent usually has an idea of the student's weaknesses and strengths. However, testing will provide a more exact knowledge of the concepts and skills that have been mastered. Some homeschooling parents also need to test and grade the student as required by their state, and these tests help them to do so.

Chapter 1 Test

Grading

My suggestion for grading the chapter 1 test is below. The total is 26 points. Divide the student's score by the total of 26 to get a decimal number, and change that decimal to percent to get the student's percentage score.

Question	Max. points	Student score
1	3 points	
2	2 points	
3	2 points	
4	3 points	
5	2 points	

Question	Max. points	Student score
6	2 points	
7	2 points	
8	4 points	
9	3 points	
10	3 points	
Total	26 points	

1. a. 56 b. 605 c. 185,725

2. Y = 28,451 (Add 8,687 and 19,764 to solve for *y*.)

3. a. 144 b. 76

4. a. 901 b. 311 b. 809

5. a. 31 b. 80

6. a. 11*s* or 11 × *s* or *s* × 11 b. 48/*b* = 8 or 48 ÷ *b* = 8

7. \$20 − (5 × \$2.50) = \$7.50. Her change was \$7.50.

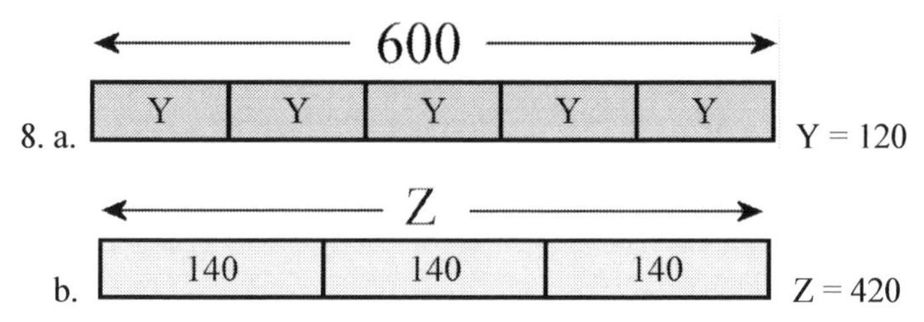

9. No, it is not. Since 990 is divisible by 3, 991 cannot be divide evenly. Or, when you add the digits of 991, you get 19, which is not divisible by 3, so 991 is not either.

10. a. 16 = 2 × 2 × 2 × 2 b. 34 = 2 × 17 c. 80 = 2 × 2 × 2 × 2 × 5

Chapter 2 Test

Grading

My suggestion for grading the chapter 2 test is below. The total is 36 points. Divide the student's score by the total of 36 to get a decimal number, and change that decimal to percent to get the student's percentage score.

Question	Max. points	Student score
1	3 points	
2	3 points	
3	12 points	
4	3 points	
5	2 points	

Question	Max. points	Student score
6	4 points	
7	2 points	
8	6 points	
9	1 point	
Total	36 points	

1. a. 70,006,324 b. 4,000,032,000 c. 98,089,000,098

2. a. 90,000 b. 30 million or 30,000,000 c. 4 billion or 4,000,000,000

3.

number	183,602	355,079,933	29,928,900
to the nearest 1,000	184,000	355,080,000	29,929,000
to the nearest 10,000	180,000	355,080,000	29,930,000
to the nearest 100,000	200,000	355,100,000	29,900,000
to the nearest million	0	355,000,000	30,000,000

4. a. 81 b. 1,000 c. 27

5. a. $6^2 = 36$ b. $2^5 = 32$

6. a. 36,000,000 b. 60,000,000 c. 70,000 d. 48,000,000

7.

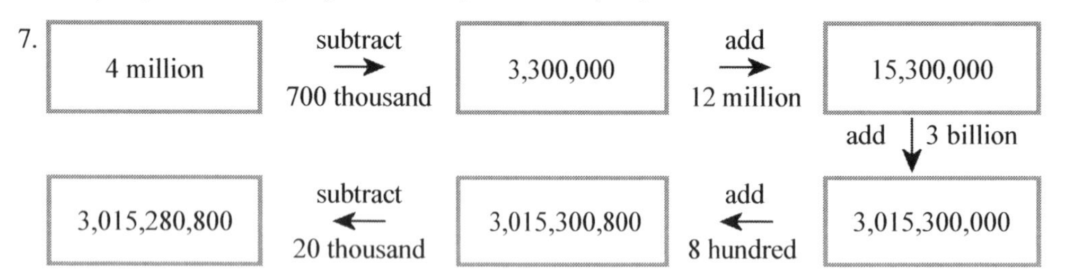

8. Estimations may vary.

a. 209,800 − 4,730	b. 2,543 × 5,187
Estimation: 210,000 − 5,000 = 205,000	Estimation: 2,500 × 5,000 = 12,500,000 Also acceptable 3,000 × 5,000 = 15,000,000
Exact answer: 205,070	Exact answer: 13,190,541
Error of estimation: 70	Error of estimation: 690,541 or 1,809,459

c. 56,493,836 + 345,399 + 7,089,400

Estimation: 56,000,000 + 345,000 + 7,000,000 = 63,345,000
Also acceptable: 56,000,000 + 0 + 7,000,000 = 63,000,000

Exact answer: 63,928,635 Error of estimation: 583,635 or 928,635

9. A family of four would have to pay $9,200 for it. ($700,000,000,000 / 305,000,000 = $2,295.08)
Hint: If the numbers do not fit onto your calculator's screen, remove the same amount of zeros from both the dividend and the divisor. Then divide. In other words, $700,000,000,000 ÷ 305,000,000 becomes $700,000 ÷ 305, and the two problems have the same answer.

Chapter 3 Test

Grading

My suggestion for grading the chapter 3 test is below. The total is 22 points. Divide the student's score by the total of 22 to get a decimal number, and change that decimal to percent to get the student's percentage score.

Question	Max. points	Student score
1	4 points	
2	4 points	
3	3 points	
4	3 points	
5	2 points	
6	3 points	
7	3 points	
Total	22 points	

1.

a. Equation: $4x = 32 + 4$	b. Equation: $x + 47 = 2x$
$(4x = 36)$	$(47 = x)$
Solution: $x = 9$	Solution: $x = 47$

2. a. $2x + 100 = 302$. Solution: $x = 101$. (Subtract 100 from 302, then divide the result by 2.)

   b. $4x + 442 = 998$. Solution: $x = 139$. (Subtract 442 from 998, then divide the result by 4.)

3. The phones cost \$120. (The discount is 1/6 of \$48, or \$8. So, one discounted phone costs \$40, and three cost \$120.)

4. The younger sister gets 109 rocks.
   First subtract $250 - 32 = 218$ to get the amount they both would have if the "extra" 32 weren't there. Then divide that by two to get 109.

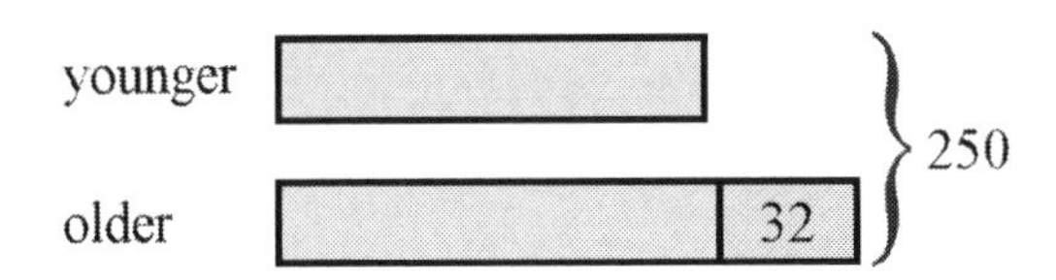

5. a. Two kilograms cost \$3. (One kilogram costs $\$7.50 \div 5 = \$1.50$.)
   b. Henry's change is \$7.

6. Since the high-quality drive costs 3 times as much as the low-quality one, the bar model has *four* parts.
   One part is $\$820 \div 4 = \$205$. The low-quality hard drive costs \$205.

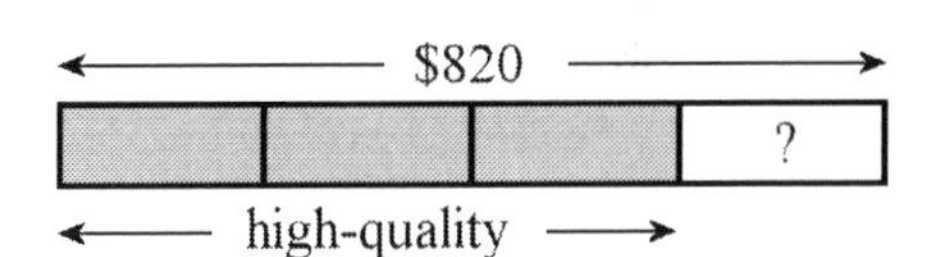

7. One block in the bar model is $66 \text{ cm} \div 3 = 22 \text{ cm}$.
   Dad is therefore $8 \times 22 \text{ cm} = 176 \text{ cm}$ tall.

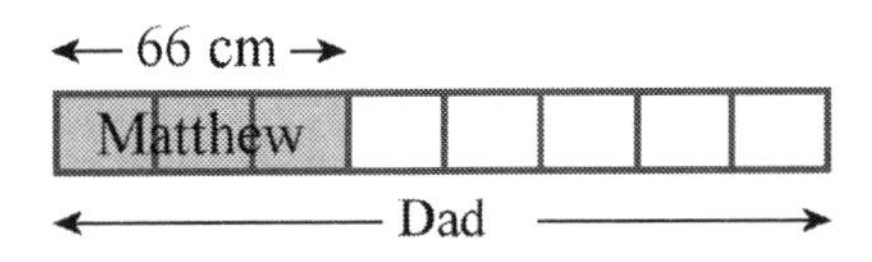

Chapter 4 Test

Grading

My suggestion for grading the chapter 3 test is below. The total is 83 points. Divide the student's score by the total of 83 to get a decimal number, and change that decimal to percent to get the student's percentage score.

Question	Max. points	Student score
1	5 points	
2	4 points	
3	4 points	
4	4 points	
5	4 points	
6	6 points	
7	6 points	
8	6 points	
9	4 points	
10	1 point	

Question	Max. points	Student score
11	4 points	
12	4 points	
13	2 points	
14	3 points	
15	6 points	
16	6 points	
17	3 points	
18	4 points	
19	3 points	
20	4 points	
Total	83 points	

1. a. 0.088 b. 0.091 c. 0.10 or 0.1 d. 0.107 e. 0.112

2. a. 5.7 b. 0.24 c. 0.35 d. 0.02

3. a. 0.21 b. 0.046 c. 3.07 d. 20.2

4. a. $\frac{6}{10}$ b. $\frac{82}{100}$ c. $1\frac{208}{1000}$ d. $\frac{93}{1000}$

5. a. < b. > c. < d. <

6.

rounded to...	nearest one	nearest tenth	nearest hundredth
8.816	9	8.8	8.82
1.495	1	1.5	1.50

rounded to...	nearest one	nearest tenth	nearest hundredth
0.398	0	0.4	0.40
9.035	9	9.0	9.04

7. a. 2.8 b. 0.63 c. 1 d. 9 e. 200 f. 0.64

8. a. 0.04 b. 0.009 c. 0.02 d. 0.08 e. 0.043 f. 0.007

9. a. 500 b. 780,000 c. 0.035 d. 1.32

10. 1.211 (The sum is 1.109 + 0.102 = 1.211)

11. a. 0.6 × 20 = 12 b. 13.08

12. a. 1.306 b. 5.25

13. Answers will vary. See some example answers below:
The answer is less than 0.8 because you multiply 0.8 by a number that is less than one.
The answer has to be less than 0.8 because multiplying by 0.9 means you are taking a fractional part (9/10) of 0.8.
Multiplying by 0.9 means taking 9/10 part of 0.8, and 9/10 is less than 1, so the answer is less than 0.8.
The answer is also less than 0.9.

14. Each box weighs 7 kg ÷ 4 = 7/4 kg = 1 3/4 kg (*or* 1.75 kg *or* 1 kg 750 g *or* 1,750 g).

15.

a. 0.7 m = 70 cm 3.2 km = 3,200 m	b. 2,650 ml = 2.65 L 0.9 L = 900 ml	c. 5.16 kg = 5,160 g 400 g = 0.4 kg

16.

a. 8 ft 10 in. = 106 in. 183 in. = 15 ft 3 in	b. 2 gal 3 C = 35 C 45 oz = 5 C 5 oz	c. 81 oz = 5 lb 1 oz 165 oz = 10 lb 5 oz

17. 1.6 L *or* 1,600 ml. (0.9 L + 350 ml + 350 ml = 0.9 L + 0.7 L = 1.6 L)

18. a. $1.12. First, find the kilogram price. Since 2 kg costs $4.48, 1 kg costs $2.24, and 1/2 kg costs $1.12.

    b. She should charge $0.56. Simply find 1/2 of the price for 1/2 kg, which was $1.12.

19. Two discounted DVDs cost $23.94. First, find the price of one discounted CD: $19.95 ÷ 5 × 3 = $11.97. Then multiply that by 2.

20. a. b.

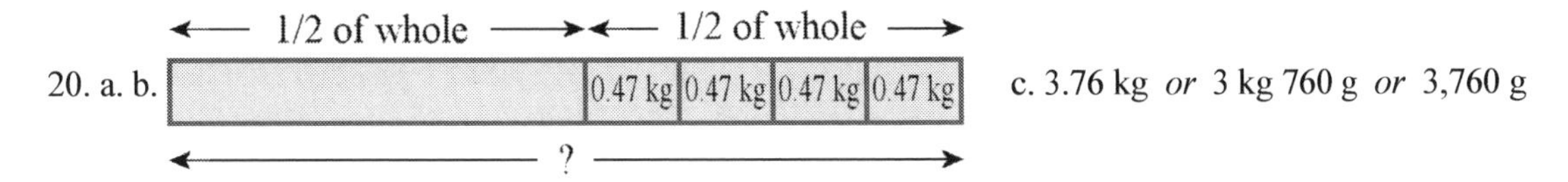

c. 3.76 kg *or* 3 kg 760 g *or* 3,760 g

Chapter 5 Test

Grading

My suggestion for grading the chapter 5 test is below. The total is 21 points. Divide the student's score by the total of 21 to get a decimal number, and change that decimal to percent to get the student's percentage score.

Question	Max. points	Student score
1	2 points	
2	4 points	
3a	1 point	
3b	2 points	
3c	1 point	

Question	Max. points	Student score
4a	4 points	
4b	1 point	
4c	1 point	
5a	2 points	
5b	1 point	
5c	1 point	
5d	1 point	
Total	21 points	

1. The coordinates of the other points are (5, 3).

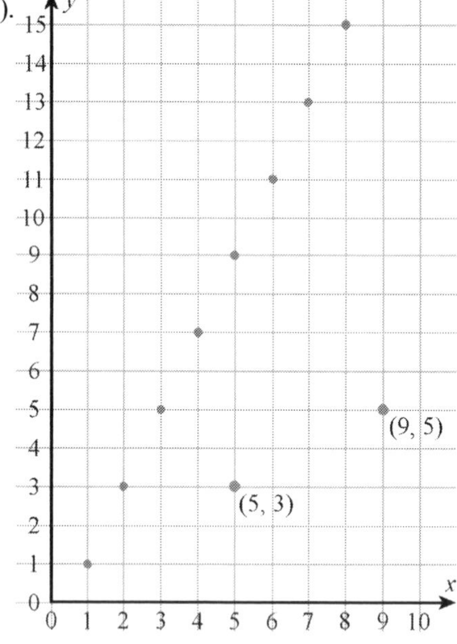

2.

x	1	2	3	4
y	1	3	5	7

x	5	6	7	8
y	9	11	13	15

3. a. b.

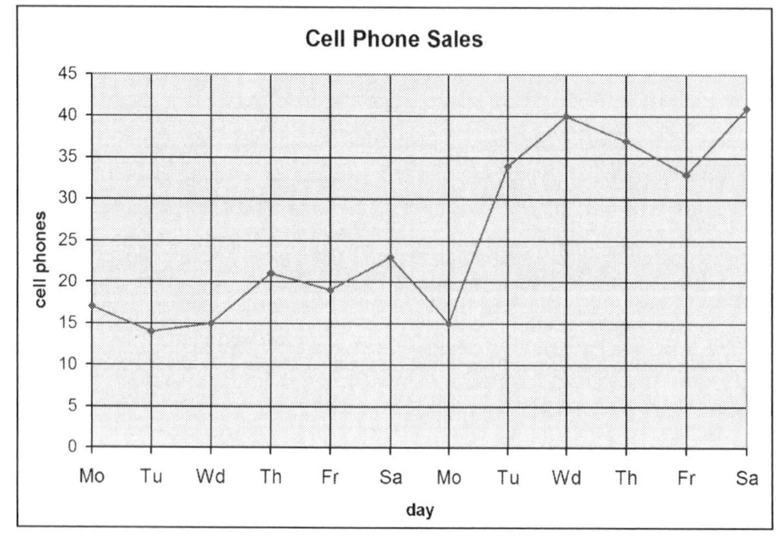

c. The promotion started on the second Tuesday.

4. a.

Age	frequency
8-9	5
10-11	8
12-13	4
14-15	3

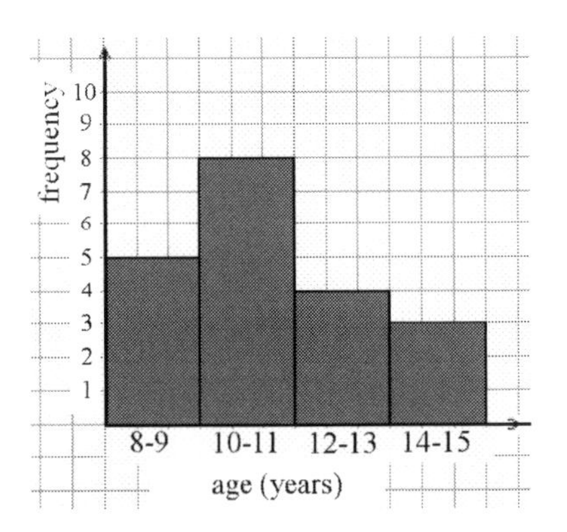

b. The mode is 10.
c. The average is 11.05.

5. a. & c.

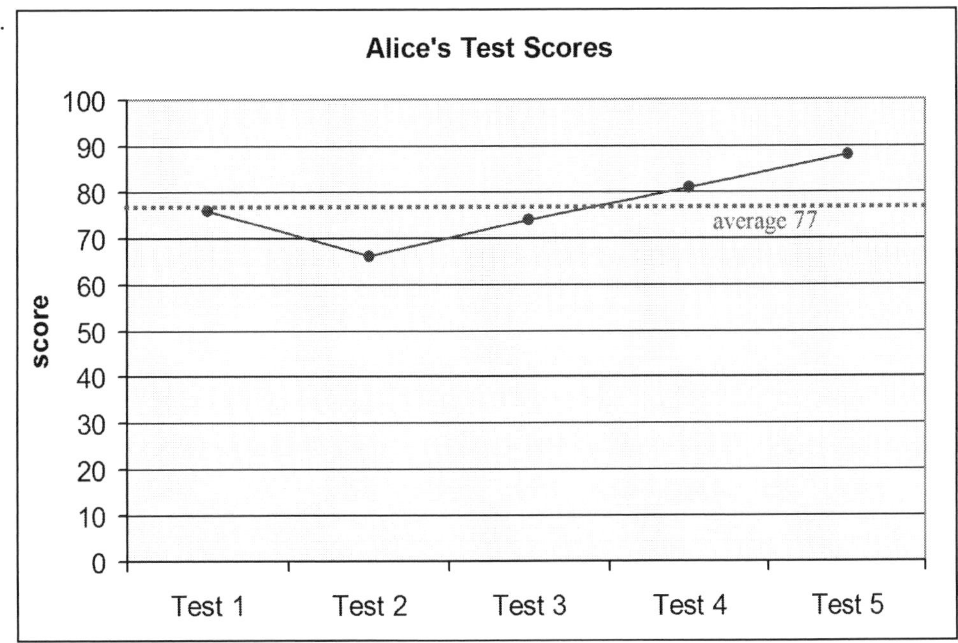

b. 77 d. 79.75

Chapter 6 Test

Grading

My suggestion for grading the chapter 6 test is below. The total is 39 points. Divide the student's score by the total of 39 to get a decimal number, and change that decimal to percent to get the student's percentage score.

Question	Max. points	Student score
1	3 points	
2	3 points	
3	3 points	
4	5 points	
5	5 points	
6	5 points	

Question	Max. points	Student score
7	3 points	
8	4 points	
9	2 points	
10	2 points	
11	4 points	
Total	39 points	

1. a. 8 2/3   b. 6 3/7   c. 6 4/5

2. a. 10 3/8   b. 2 2/5   c. 16 8/11

3. a. 2 5/7   b. 2 7/9   c. 4 9/15

4.

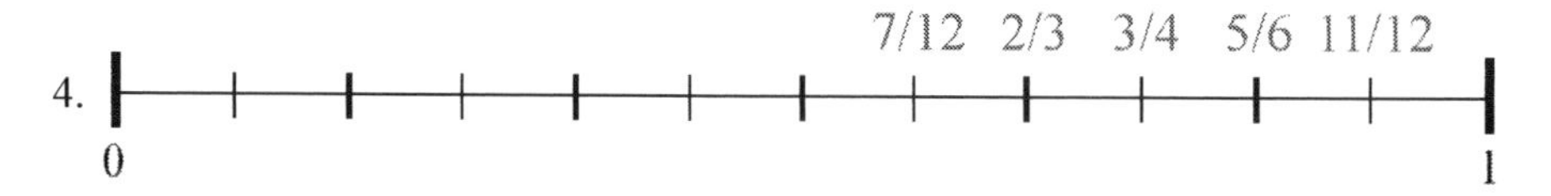

5.

a. $\frac{3}{7} = \frac{9}{21}$	b. $\frac{4}{3} = \frac{24}{18}$	c. $\frac{5}{6} = \frac{\quad}{11}$ NOT POSSIBLE	d. $\frac{2}{5} = \frac{8}{20}$	e. $\frac{5}{6} = \frac{15}{18}$

6.

a. $\frac{7}{4} > \frac{5}{3}$	b. $\frac{5}{11} < \frac{1}{2}$	c. $\frac{7}{10} > \frac{69}{100}$	d. $\frac{3}{4} = \frac{75}{100}$	e. $\frac{8}{7} > \frac{7}{9}$

7. We split 1/3 into additional pieces so it becomes 5/15, and similarly, we split 2/5 into additional pieces so they become 6/15. Now we can add. The answer is 11/15.

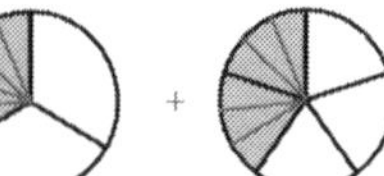

8. a. 1 5/12   b. 1/6   c. 2 9/14   d. 10 3/40

9. $\frac{1}{2}, \frac{5}{9}, \frac{4}{7}, \frac{7}{5}$

10. 1/4 + 21/28 = 1 *or* 2/4 + 14/28 = 1 *or* 3/4 + 7/28 = 1

11. The sides measure 3 3/16 inches., 1 5/16 inches., and 2 1/4 inches. The perimeter is 6 3/4 inches.

Chapter 7 Test

Grading

My suggestion for grading the chapter 7 test is below. The total is 36 points. Divide the student's score by the total of 36 to get a decimal number, and change that decimal to percent to get the student's percentage score.

Question	Max. points	Student score
1	2 points	
2	3 points	
3	2 points	
4	3 points	
5	3 points	
6	6 points	
7	4 points	

Question	Max. points	Student score
8	2 points	
9	2 points	
10	2 points	
11	3 points	
12	2 points	
13	2 points	
Total	36 points	

1\. $\frac{6}{10} = \frac{3}{5}$ (÷ 2 in the numerator and ÷ 2 in the denominator)

The parts were joined together in twos .

2\. a. 3 2/3 b. 2/3 c. 1 3/32

3\. She would need 5 × (2/3 C) = 3 1/3 cups of butter for 5 batches of cookies.

4\. Draw five copies of 3/4: . This equals or 3 3/4.

5\. It is not. The illustration shows (3/4) × (1/2) = (1/3), but in reality (3/4) × (1/2) = 3/8.

6\.

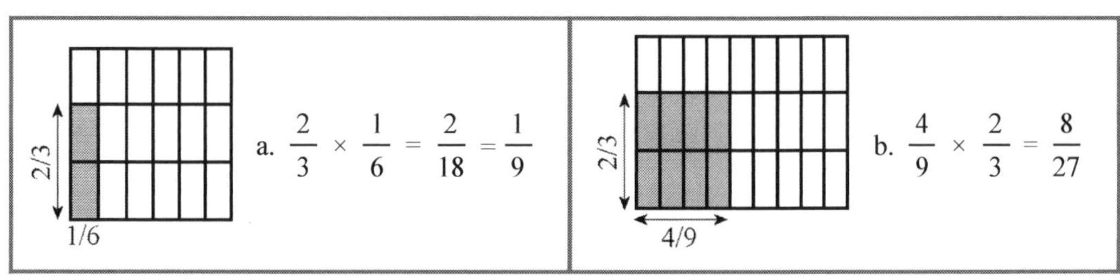

7\. a. 5/18 b. 7 1/5

8\. The area is 3 33/64 square inches.

9\. Each person had 1/9 of the original pizza. (1/3) ÷ 3 = 1/9.

10\. a. You can get nine servings. b. 3 ÷ (1/3) = 9

11\. a. 1/18 b. 48 c. 3/11

12\. Instead of having 3 hearts, 2 circles, and 2 diamonds, a drawing could have double or triple as many of each shape. The ratio is 3:2:2.

13\. One block in the model is 80. There are 160 white marbles.

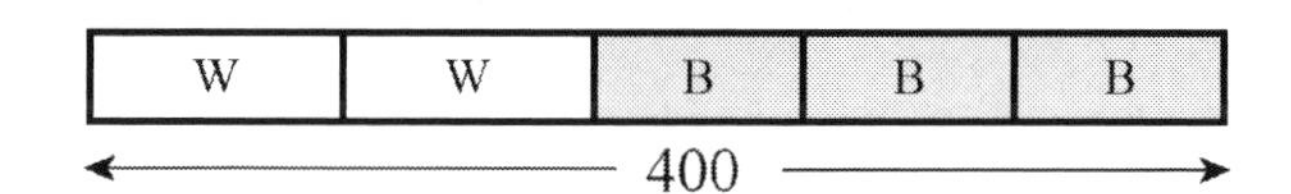

Chapter 8 Test

Grading

My suggestion for grading the chapter 8 test is below. The total is 32 points. Divide the student's score by the total of 32 to get a decimal number, and change that decimal to percent to get the student's percentage score.

Question	Max. points	Student score
1	5 points	
2	2 points	
3a	2 points	
3b	2 points	
3c	2 points	
3d	2 points	
4	3 points	

Question	Max. points	Student score
5	4 points	
6	1 point	
7	2 points	
8	3 points	
9a	3 points	
9b	1 point	
Total	32 points	

1. a. rhombus b. trapezoid c. right scalene triangle d. isosceles obtuse triangle e. kite

2. Yes, it is a kite, because it has four congruent sides it is a rhombus, and all rhombi are also kites.
   Yes, it is a trapezoid, because all parallelograms are trapezoids. It may or may not be a square (we do not have enough information).

3. a. Yes, it is. A kite has two pairs of congruent sides, and the congruent sides are adjacent. In a square, all sides are congruent, so that fulfills the definition of a kite.

   b. Yes, it is, because it has two sides that are parallel.

   c. Yes, it can, if it is a square. Check the student's sketches; they should show a square.

   d. No, it cannot. An equilateral triangle has three angles that are 60°; therefore it does not have any right angles.

4. a. perimeter b. volume c. area

5. Student's triangles vary, but should have the same basic shape as the example triangle on the right. Start out by drawing the base side. Then draw the 30°-angles and continue the sides until they meet. The top angle measures 120°.

6. The area is 1 square inch.

7. The volume is 2 ft × 1.5 ft × 1.5 ft = 4.5 cubic feet.

8. a. The volume of one book is 15 cm × 30 cm × 1.5 cm = 675 cm^3.
   b. The volume of six books is 6 × 675 cm^3 = 4,050 cm^3.

9. The triangle is acute and scalene.

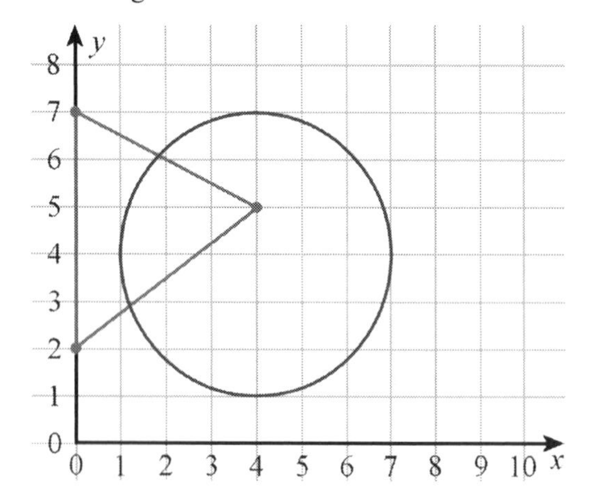

The End of the Year Test

Please see the file for the End of the Year Test for grading instructions.

The Four Operations

1. a. 45 b. 409,344

2. a. x = 296,430 b. Y = 80 c. N = 3,304

3. All of these are correct:
$4Y = 600$ or $4 \times Y = 600$ or $Y + Y + Y + Y = 600$ or $600 \div 4 = Y$ or $600 \div Y = 4$ or $600 - Y - Y - Y - Y = 0$.
Solution: Y = 150.

4. a. $42 \times 10 = (10 - 4) \times 70$ b. $143 = 13 \times (5 + 6)$

5. (\$19.95 − \$5) × 5 or 5 × (\$19.95 − \$5). Her total cost was \$74.75.

6. No, it is not. Explanations vary. For example: It is an odd number, and therefore cannot be divisible by an even number.
991 ÷ 4 = 247 R3, leaving a remainder, so 991 is not divisible by 4.

7. a. $26 = 2 \times 13$ b. $40 = 2 \times 2 \times 2 \times 5$ c. 59 is prime

Large Numbers

8. a. 70,016,090 b. 32,000,232,000

9. It is about 32,000 × 300 = 9,600,000. Other estimates are also possible.

10. 80 million or 80,000,000

11.

number	593,204	19,054,947
to the nearest 1,000	593,000	19,055,000
to the nearest 10,000	590,000	19,050,000
to the nearest 100,000	600,000	19,100,000
to the nearest million	1,000,000	19,000,000

Problem Solving

12. An 8-ft long board is 96 inches. One-sixth of that is 96 in. ÷ 6 = 16 in. The remaining piece is 80 inches, or 6 ft 8 in.

13. It would cost \$7.80 to download ten songs. First, find the price of one song download: \$4.68 ÷ 6 = \$0.78. Then, multiply that by 10.

14. A lunch in the cheap restaurant costs 1/3 of \$36, or \$12. Mary spends \$36 + 4 × \$12 = \$84.

15.

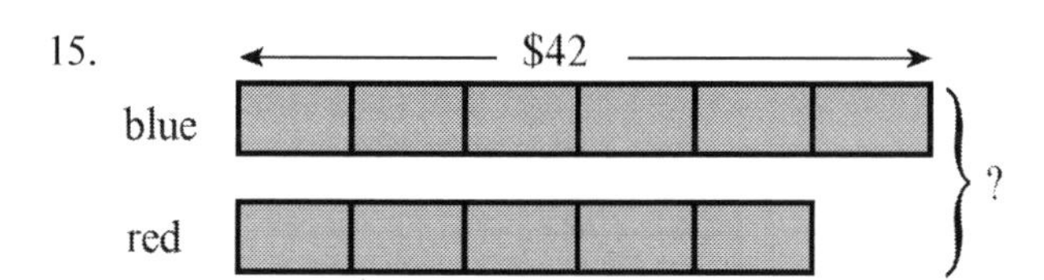

One block in the model is $42 ÷ 6 = $7. The red swimsuit costs 5 × $7 = $35. Together they cost $77.

16. a.

← 134 →

green	green	purple	purple	purple

b. One block or part in the model is 134 ÷ 2 = 67 marbles. There are 3 × 67 = 201 purple marbles.

17. a. The DVD costs about $30. Karen pays 3/5 of it, which is about $30 ÷ 5 × 3 = $18. Ann pays about $12.

b. Karen pays $29.90 ÷ 5 × 3 = $17.94. Ann pays $11.96.

Decimals

18. a. 0.289 b. 0.30 c. 0.305 d. 0.313

19. a. 0.95 b. 0.72 c. 0.62 d. 1.26 e. 1.05 f. 0.37

20. a. 0.08 b. 0.081 c. 5.21

21. a. $\frac{48}{1000}$ b. $1\frac{4}{1000}$ c. $7\frac{22}{100}$

22. a. 0.31 > 0.031 b. 0.43 > 0.093 c. 1.6 > 1.29

23.

rounded to...	nearest one	nearest tenth	nearest hundredth
5.098	5	5.1	5.10

rounded to...	nearest one	nearest tenth	nearest hundredth
0.306	0	0.3	0.31

24.

a. 0.4 × 7 = 2.8 b. 0.4 × 0.7 = 0.28 c. 0.4 × 700 = 280	d. 10 × 0.05 = 0.5 e. 100 × 0.05 = 5 f. 1000 × 0.5 = 500	g. 1.1 × 0.3 = 0.33 h. 70 × 0.9 = 63 i. 20 × 0.09 = 0.18

25.

a. 0.36 ÷ 6 = 0.06 b. 5.6 ÷ 7 = 0.8	c. 3 ÷ 100 = 0.03 d. 0.7 ÷ 10 = 0.07	e. 16 ÷ 10 = 1.6 f. 71 ÷ 100 = 0.71

26.

a. 0.2 m = 20 cm 37 cm = 0.37 m 2.9 km = 2,900 m	b. 0.4 L = 400 ml 3.5 kg = 3,500 g 240 g = 0.24 kg	c. 56 oz = 3 lb 8 oz 74 in. = 6 ft 2 in. 15 C = 3 qt 3 C

27. There are 444 milliliters in two bowls. Two liters is 2,000 ml. 2,000 ml ÷ 9 = 222.2 ml or about 222 ml.

28. a. 1.42 b. 14.28 b. 14.08

Graphs

29.

x	0	1	2	3	4	5
y	1	3	5	7	9	11

30. See the image on the right.

31.

Day	Sales (1000 dollars)
Mon	125
Tue	114
Wed	118
Thu	130
Fri	158

a. See the line graph on the right.

b. The average daily sales is $129,000.

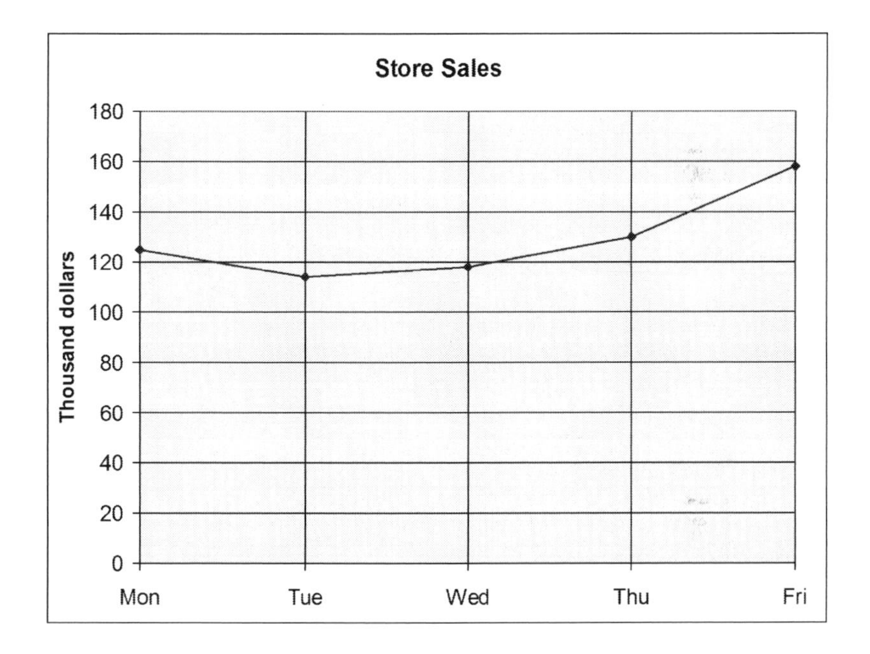

Fractions

32. a. 6 1/3 b. 2 1/3 c. 13 4/5

33. Number line: 0, 1/3, 5/12, 4/6, 3/4, 1

34.

~~a. $\frac{5}{6} = \frac{\ }{20}$~~	b. $\frac{2}{7} = \frac{8}{28}$	c. $\frac{3}{8} = \frac{15}{40}$	d. $\frac{2}{9} = \frac{6}{27}$

35. Mia found the common denominator (15) correctly, but forgot that the 2 fifths and the 2 thirds do not stay as 2 fifteenths in the conversion.

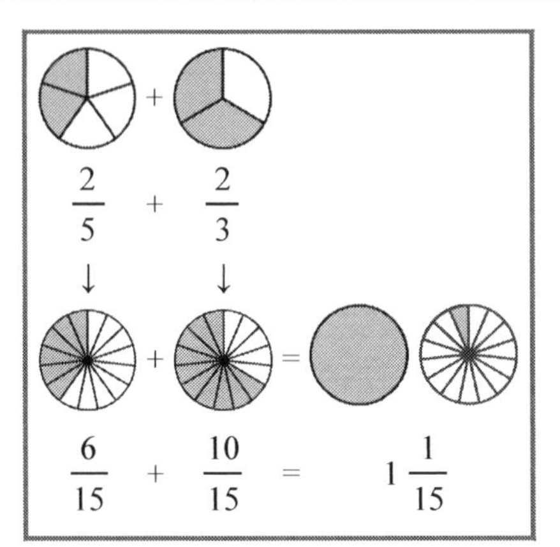

36. 1 1/6 b. 7/15 c. 5 5/8 d. 10 5/18

37. You would need 3 × (2 3/4) = 8 1/4 cups of flour to make three batches of rolls.

38. a. $\frac{6}{9} > \frac{6}{13}$ b. $\frac{6}{13} < \frac{1}{2}$ c. $\frac{5}{10} > \frac{48}{100}$ d. $\frac{1}{4} = \frac{25}{100}$ e. $\frac{5}{7} > \frac{7}{10}$

39. a. 1 2/5 b. cannot be simplified c. 7/8

40. Yes, it is correct. (2/3) × (1/2) = 1/3.

41.

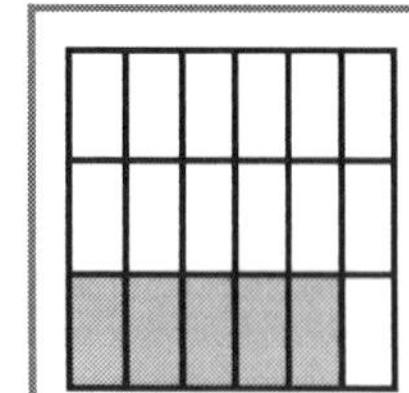

a. $\frac{1}{3} \times \frac{5}{6} = \frac{15}{18}$

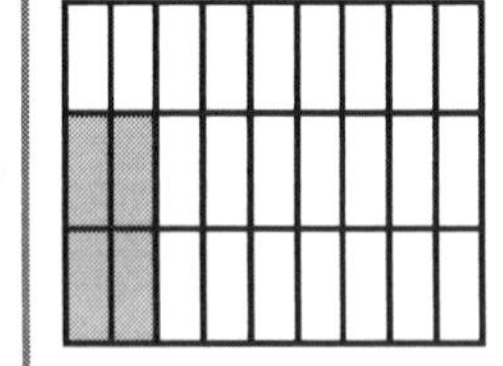

b. $\frac{2}{9} \times \frac{2}{3} = \frac{4}{27}$

42. You can cut 60 pieces. 15 in. ÷ (1/4 in.) = 60

43. 1/6 of the pizza. (1/2) ÷ 3 = 1/6

44. a. 10 1/2 b. 1/21 c. 2 14/15 d. 18

Geometry

45. The sides measure 2 15/16 in., 2 9/16 in., and 4 15/16 in. The perimeter is 10 7/16 in.

46. a. an isosceles acute triangle b. a rhombus c. a right scalene triangle d. a trapezoid

47. a. 9 m^2 b. 20 ft

48. Yes, it is. A square has one pair of parallel sides, which is a definition of a trapezoid.

49. Yes, it can. For example

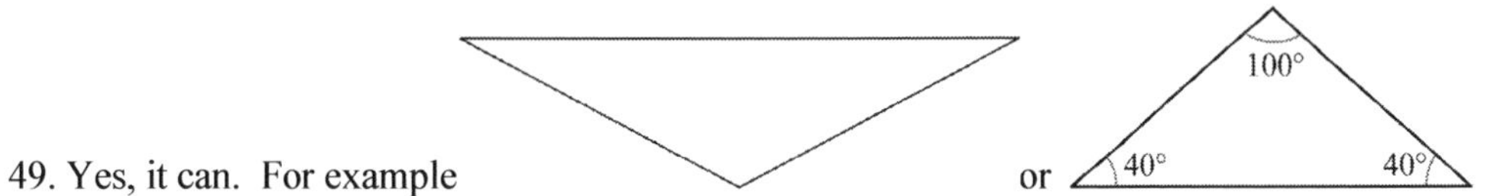

50. a. Check the student's triangles. The student should use a tool, such as a triangular ruler or a protractor, to make the right angle. The picture below may be slightly out of scale when printed, due to the possible scaling in the printing process.

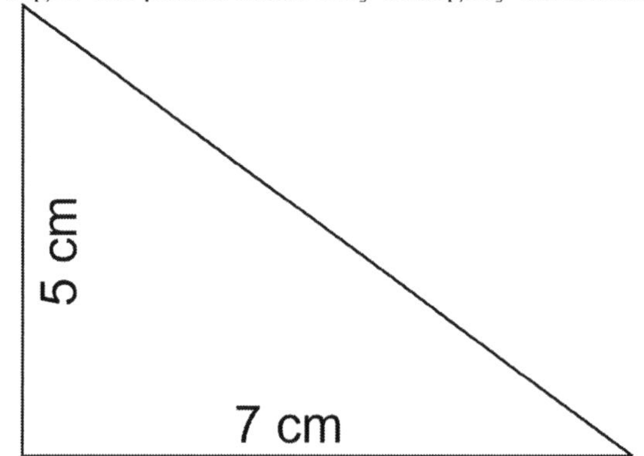

b. 8.6 cm + 5 cm + 7 cm = 20.6 cm
c. They measure _90_ °, _36_ °, and _54_ °.

51. The volume is 5 cm × 10 cm × 4 cm = 200 cm^3.

52. a. 1.2 m × 0.6 m × 1 m = 0.72 m^3.
b. 240 liters. 0.72 m^3 is 720 liters, and one-third of that is 240 liters.

Cumulative Reviews Answer Keys

Cumulative Reviews Answer Key, Grade 5

Cumulative Review: Chapters 1 - 2

1. a. $90 + (70 + 80) \times 2 = 390$ b. $378 = 6 \times (8 + 13) \times 3$ c. $90 \times 4 = (180 - 60) \times 3$

2. a. x = 20

←— 200 —→				
x	x	x	x	120

b. x = 9

←— 52 —→			
25	x	x	x

3. a. 630 b. 322

4. a. He earned $480 in one week. b. They raced 1,496 miles.

5. **(3)** $50 – ($9 + $9 + $9 + $9) and **(5)** $50 – 4 × $9 = $14. His change was $14.

6. Estimates may vary. The method I use most often in estimating multiplications is to round one number up, the other down, to numbers that are easy to multiply mentally. For example, in 173 × 35, rounding to 200 × 30 may look "unorthodox", but it gives a good estimate.

a.	b.	c.
$\begin{array}{r} 173 \\ \times\ 35 \\ \hline 865 \\ 5190 \\ \hline 6055 \end{array}$	$\begin{array}{r} 269 \\ \times\ 537 \\ \hline 1883 \\ 8070 \\ 134500 \\ \hline 144453 \end{array}$	$\begin{array}{r} 892 \\ \times\ 340 \\ \hline 0 \\ 35680 \\ 267600 \\ \hline 303280 \end{array}$
Estimate: 200 × 30 = 6,000	Estimate: 300 × 500 = 150,000	Estimate: 900 × 300 = 270,000
Error of estimation: 55	Error of estimation: 5,547	Error of estimation: 33,280

Cumulative Review: Chapters 1 - 3

1. 389

2. a. 124 b. 84 c. 55 d. 490

3. a. 26, 28 b. 47, 135 c. 450, 600

4. a. 78,000,016,038 b. 844,012,000,704

5.

number	32,274,302	64,321,973	388,491,562	2,506,811,739
to the nearest 1,000	32,274,000	64,322,000	388,492,000	2,506,812,000
to the nearest 10,000	32,270,000	64,320,000	388,490,000	2,506,810,000
to the nearest 100,000	32,300,000	64,300,000	388,500,000	2,506,800,000
to the nearest million	32,000,000	64,000,000	388,000,000	2,507,000,000

6.

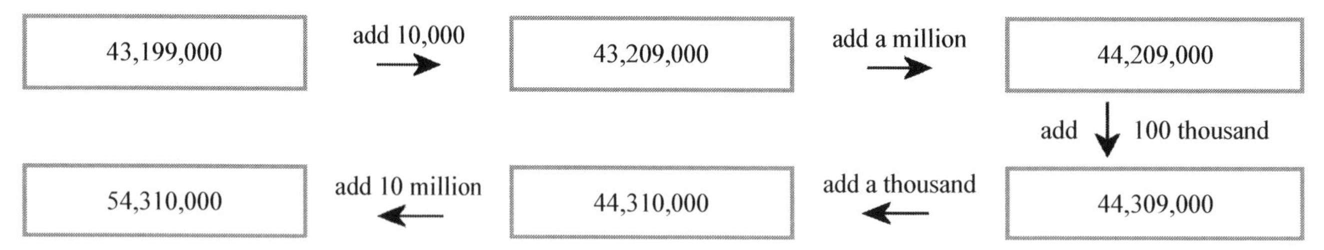

7. a. $5 \times 6 + 50$ b. $10 - (9 - 6)$

8. $(21 \times 2) + (20 \times 1.50) + 12 =$ \$84 total cost.

9. a. 4,815,598,182 b. 1,046,556,957

10. Estimates may vary.

a. $2,933 \times 213$	b. $152 \times 89 \times 7,932$
My estimation: $3,000 \times 200 = 600,000$ Exact answer: 624,729 Error of estimation: 24,729	My estimation: $150 \times 100 \times 8,000 = 120,000,000$ Exact answer: 107,304,096 Error of estimation: 12,695,904

Cumulative Review: Chapters 1 - 4

1. Together they earned \$225. Jack's sister earns $\$125 \div 5 \times 4 = \$100$.

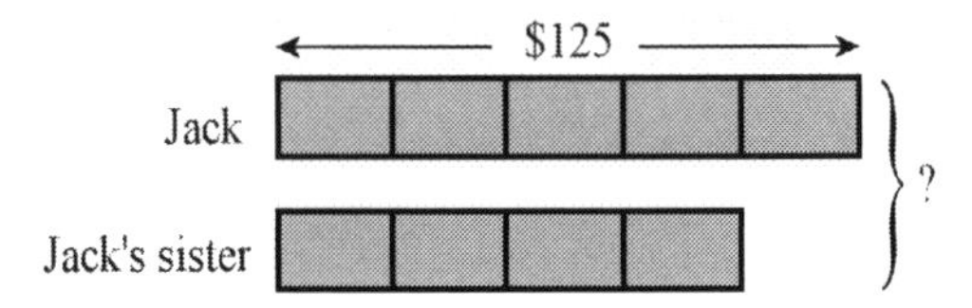

2. The price difference is \$2.60.
   The cheaper thermometer costs $\$10.40 \div 4 \times 3 = \$7.80$. Then the difference is $\$10.40 - \$7.80 = \$2.60$.
   You can also solve this by thinking that the difference in prices is 1/4 of the price of the expensive thermometer, and $\$10.40 \div 4 = \$2.60$.

3. $120 \div 5 \times 2 = 48$; $120 - 48 = 72$. She has 72 more marbles.

4. a. Eighteen students are new. Simply find 6/50 of 150: 1/50 of 150 is 3, so 6/50 is six times as much, or 18.
   b. Twelve students have studied English before. First find 1/3 of 18 students, which is 6.
      The rest, or 12, have studied English before.

5. 182

6. a. M = 50   b. M = 48   c. M = 90   d. N = 8,000   e. N = 21,000   f. N = 80

7. a. $350 \div x = 5$ b. $(15 - 6) + 8$ or $8 + (15 - 6)$ or the same expressions without parentheses.

8. a. 3   b. 11   c. 4

9.

number	**97,302**	**25,096,199**	**709,383,121**	**89,534,890,066**
to the nearest 1,000	97,000	25,096,000	709,383,000	89,534,890,000
to the nearest 10,000	100,000	25,100,000	709,380,000	89,534,890,000
to the nearest 100,000	100,000	25,100,000	709,400,000	89,534,900,000
to the nearest million	0	25,000,000	709,000,000	89,535,000,000

Cumulative Review: Chapters 1 - 5

1.

Round this to the nearest →	unit (one)	tenth	hundredth
4.925	5	4.9	4.93
6.469	6	6.5	6.47

Round this to the nearest →	unit (one)	tenth	hundredth
5.992	6	6.0	5.99
9.809	10	9.8	9.81

2. $56 \times 2 - 14 = 98$

3. a. 0.37 b. 0.192 c. 0.328 d. 1.45 e. 1.05 f. 0.506

4. a. 3.09 b. 8.075

5.

a. 0.5 m = 50 cm 0.06 m = 6 cm 2.2 km = 2,200 m	b. 4.2 L = 4,200 mL 400 mL = 0.4 L 5,400 g = 5.4 kg	c. 800 g = 0.8 kg 4,550 m = 4.55 km 2.88 kg = 2880 g

6. One package of AAA batteries costs $1.90. To solve this, first subtract the cost of the AA batteries from the total: $17.04 − $5.64 = $11.40. Then divide that by six: $11.40 ÷ 6 = $1.90.

7. a. 5 R1, 5.17 b. 10 R3, 10.75

8. a. Estimate: $2 \times 12 = 24$. Exact: 24.035
b. Estimate: $70 \times 2 = 140$. Exact: 156.22
c. Estimate: $7 \times 3 = 21$. Exact: 21.442

9. a. One packet of seeds cost $1.89 b. One plant cost $1.92.
c. The total cost was $26.67.

10. a. 0.7 b. 63 c. 29 d. 0.08 e. 0.045 f. 0.076

Cumulative Review: Chapters 1 - 6

1. Estimates vary.
   a. Estimate 300 × 280 = 84,000. Exact 80,330.
   b. Estimate 530 × 400 = 212,000. Exact 218,400.
   c. Estimate 900 × 200 = 180,000. Exact 166,842.

2. 2,400 m − 2 × 250 m = 4,300 m or 4.3 km

3. 100 − 10 ÷ 2 = $45; Angi got $55 and Rebekkah got $45.

4. a. 0.4 b. 1.2 c. 30 d. 0.2 e. 20 f. 4

5. a. $y = 1.67$ b. $z = 1.681$

6. The mean is 18.

7. a. 0.908 = 9 × (1/10) + 0 × (1/100) + 8 × (1/1000) b. 543.2 = 5 × 100 + 4 × 10 + 3 × 1 + 2 × (1/10)

8. a. 0.4 b. 0.06 c. 5 d. 0.04 e. 9 f. 20

9. Each student's share was $2,676.

10. A half of a gallon is eight cups, so there are six cups of milk left.

11. Sixty-one inches is five feet and one inch, so Eva is five inches taller than Ava.

12. 36 × 6 oz = 216 oz; 216 oz ÷ 16 oz = 13 1/2. The total weight is 13 lb 8 oz.

13. a. The histograms can vary based on how the bins are chosen. The example answer below uses a bin width of 10, starting at 0. One could also use a bin width of 11, starting at 4, or some other possibilities.

distance	frequency
0...9	5
10...19	6
20...29	3
30...39	1
40...49	1

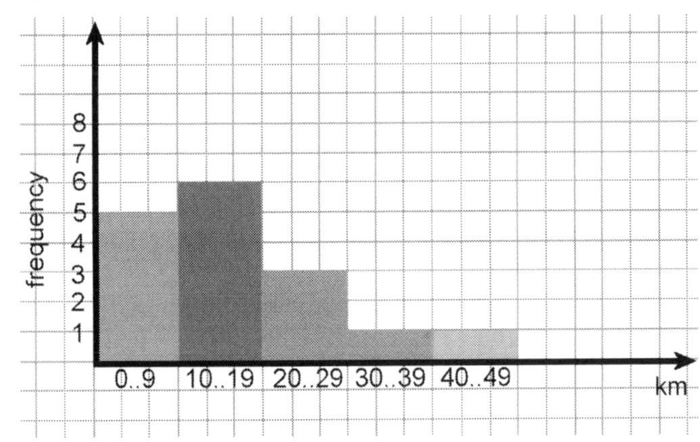

b. 17.1875 km c. There are three modes: 7 km, 15 km, and 25 km.

Cumulative Review: Chapters 1 - 7

1.

a. $13 \times 4 + 18 = 70$ $4 + 8 \div 8 = 5$	b. $(2 + 60 \div 4) \times 3 = 51$ $2 + 30 \times (7 + 8) = 452$	c. $10 \times (9 + 18) \div 3 = 90$ $5 \times (200 - 190 + 40) = 250$

2. a. One hundred apples cost $23.00   b. Ten apples in each small bag is worth $2.30.

3. a. >   b. <   c. >   d. =

4.

a. 791,456,030 Place: ten thousands place Value: fifty thousand or 50,000	b. 2,094,806,391 Place: one millions place Value: four million or 4,000,000

5.

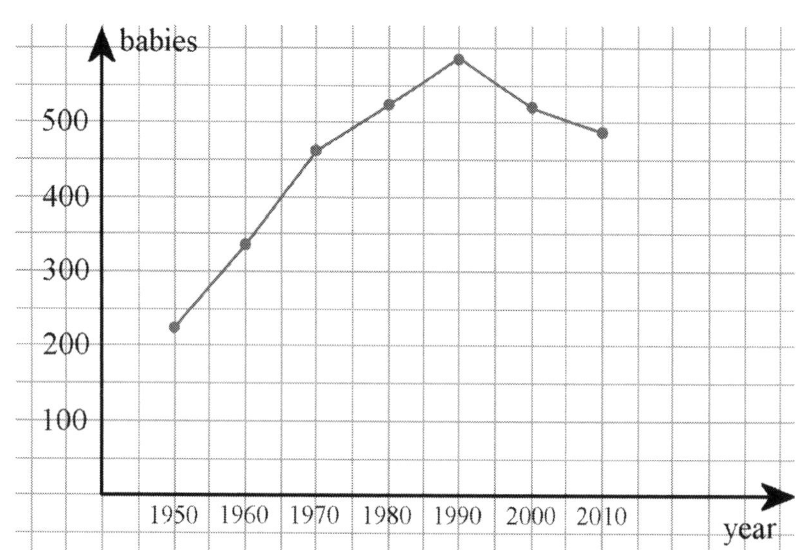

6. Equation: $4x + 176 = 516$ or $516 - 176 = 4x$ or $516 - 4x = 176$.
   Solution: $x = (516 - 176) \div 4 = 85$

7. $5 \times 1.2 = 6$ km

8. a. The estimated cost would be 8 × $1 + 6 × $1 = $14.
   b. Shelly's original bill is 8 × $1.19 + 6 × $0.88 = $14.80. Shelly pays 4/5 of it: $14.80 ÷ 5 × 4 = $11.84
      Or, you can multiply 0.8 × $14.80 = $11.84.

9. a. 3 8/55   b. 8 5/14   c. 6 13/20   d. 6 7/30

10. a. 1.13   b. 2.08   c. 8.94

11. a. 30 R1, 5.17   b. 10 R3, 10.75

Cumulative Review: Chapters 1 - 8

1.

Picture or Diagram	As Fractions	As a Ratio
	1/6 of the shapes are hearts. 5/6 of the shapes are diamonds.	The ratio of hearts to diamonds is 1:5.

2. a. 1 : 3 b. 3 : 6

3. 24 were oatmeal cookies.

4. 33 cubic inches

5. The estimates may vary.
   a. Estimate: $22 \times 4 = 88$ Exact: 84.63
   b. Estimate: $0.5 \times 1 = 0.5$ Exact: 0.416
   c. Estimate: $140 \times 5 = 700$ Exact: 736.02

6. a. 91.5 km b. $40 \times 5 \times 91.5$ km = 18,300 km

7. a. kite b. rectangle c. trapezoid d. parallelogram

8. The only way to draw this is if the two sides that are 6 cm long are the ones that meet in a right angle. (The image is not to scale.)

6 cm
8.49 cm
6 cm

9. a. 1/6 b. 1 10/11

10. You can fill ten glasses.

11. a.

AGE (yrs)	WEIGHT (kg)	Weight gain from previous year
0	3.3 kg	-
1	10.2 kg	6.9 kg
2	12.3 kg	2.1 kg
3	14.6 kg	2.3 kg
4	16.7 kg	2.1 kg
5	18.7 kg	2.0 kg
6	20.7 kg	2.0 kg
7	22.9 kg	2.2 kg
8	25.3 kg	2.4 kg
9	28.1 kg	2.8 kg

AGE (yrs)	WEIGHT (kg)	Weight gain from previous year
10	31.4 kg	3.3 kg
11	32.4 kg	1.0 kg
12	37.0 kg	4.6 kg
13	40.9 kg	3.9 kg
14	47.0 kg	6.1 kg
15	52.6 kg	5.6 kg
16	58.0 kg	5.4 kg
17	62.7 kg	4.7 kg
18	65.0 kg	2.3 kg

b. He gained the fastest at ages 1 year, 14 years, 15 years, and 16 years.
c. You can see that by how steeply the line is rising on the graph.

Made in the USA
Charleston, SC
22 June 2013